AF696722

Globalization:

Technology, Literacy & Curriculum

O'Connor-Petruso, S.A. Girelli-Carasi, F.
Co-Editors

Custom Publishing

New York Boston San Francisco
London Toronto Sydney Tokyo Singapore Madrid
Mexico City Munich Paris Cape Town Hong Kong Montreal

Copyright © 2010 by Sharon O'Connor-Petruso and Fabio Girelli-Carasi
All rights reserved.

Permission in writing must be obtained from the publisher before any part of this work may be reproduced or transmitted in any form or by any means, electronic or mechanical, including photocopying and recording, or by any information storage or retrieval system.

All trademarks, service marks, registered trademarks, and registered service marks are the property of their respective owners and are used herein for identification purposes only.

Printed in the United States of America

5

2008220187

MP

Pearson Custom Publishing is a division of

PEARSON

www.pearsonhighered.com

ISBN 10: 0-558-36225-7
ISBN 13: 978-0-558-36225-6

Globalization: Technology, Literacy, & Curricula

O'Connor-Petruso, S.A. & Girelli-Carasi, F. Co-Editors

Introduction

Sharon Anne O'Connor-Petruso

The following textbook highlights the dramatic transformations sweeping our educational system due to the second generation of the Internet, Web 2.0, and the critical need for professors and teachers to train their students in new and emerging technologies regardless of limited funding, outdated infrastructures, and narrow shelf life of applications. Successful step-by-step instructional methodologies in utilizing open source software, including free web hosting, in a variety of courses is provided. Elicited by globalization, the new meaning of literacy and strategies to prepare urban minority youth to become productive citizens in the global economy are discussed.

Educational systems in combination with government and industry leaders worldwide (eEurope 2005 Action Plan, 2003; i2010-Information Space Innovation & Innovation in R & D Inclusion, 2005; Intel@Teach to the Future, 2005; Toward A New Golden Age In American Education: How The Internet, The Law And Today's Students Are Revolutionizing Expectations, 2005; UNESCO, 2003) are creating rich technology learning experiences that will help produce competitive and productive citizens (Dewey, 1917) who are work ready for the global economy. Thus it is critical that today's students be exposed, trained, and have ongoing participation with the digital tools of Web 2.0

including shareware and free courseware which attempts to abate the growing Digital Divide. However students are dependent upon the Ed Tech Policies, plans, visions, assessment strategies, and procedures and training their educational system employs and how they address the ever growing dilemma of outdated infrastructures and limited shelf life of applications (O'Connor-Petruso & Lei, in press).

In addition to Information Age technology skills, globalization demands literate workers who are multilingual and at a bare minimum competent in the universal language of English. The latest National Assessment of Educational Progress (2007) reported that American twelfth grade reading scores are the lowest in twenty years.

Thus this textbook highlights various perspectives on how the faculty of the School of Education and Department of Modern Languages and Literacy are rising to the challenge to produce technologically savvy professors who can turnkey Information Age skills to teacher candidates and teachers who are literate and espouse the tenets of diversity, collaboration, social justice, and reflection – critical qualities necessary for successful globalization.

Chapter 1:

Embedding Asynchronous and Synchronous Technologies in Course Curricula

Sharon Anne O'Connor-Petruso

Brooklyn College of the City University of New York

oconnorpetrusos@brooklyn.cuny.edu

Electronic learning, e-learning, is transforming how we both teach and learn. Albeit today's educational infrastructures support Internet capability, majority of the e-learning and use of digital tools is more home-based than school-based (Visions 2020.2, n.d.). Thus it is critical for professors to model/embed asynchronous and synchronous technologies into their course curricula so teachers and teacher candidates can produce globally competitive citizens who are competent in new and emerging technologies. This chapter focuses upon popular Web 2.0 tools and freeware (in response to the growing Digital Divide) as the less economically advantaged will not have access to pricey advanced hardware and software.

Chapter 2:

Instructional Strategies for Integrating Technology into the Curriculum

Barbara Rosenfeld

Brooklyn College of the City University of New York

Rosenfeld@brooklyn.cuny.edu

The use of any instructional support materials must be coherent with the lesson objective. This is even more important with multimedia and information technology. The teacher and the students must avoid the appeal of the fun and playful aspects to concentrate instead on a purposeful rationale for classroom activities. This chapter outlines effective processes and procedures for the integration of technology with a holistic, omni-comprehensive perspective, beginning from the specific goals of a single lesson to the end objectives of contributing to digital literacy and professional growth and leadership.

Chapter 3:

Literacy – A Critical Constituent for Successful Globalization

Meral Kaya, Sharon Anne O'Connor-Petruso, & Fabio Girelli-Carasi

Brooklyn College of the City University of New York

mkaya@brooklyn.cuny.edu, oconnorpetrusos@brooklyn.cuny.edu,

& carasi@brooklyn.cuny.edu

Teaching literacy carries the responsibility of raising literate "GLOBAL" citizens which include competencies in the universal language of English, mathematics, and technology and adept in visual, communication, and calculation skills (Cooper & Kiger, 2006; Goldberg, 2006). This new broad definition of literacy also brings changes in literacy teaching methods, approaches, and instructional materials. This chapter will explore various teaching ideas and tools of technology that enhance literacy teaching and learning as well as "hands-on" activities and assignments that provide more authentic assessment.

Chapter 4:

Globalization, Technology and the Education of Young Children: Understanding the World in Two Languages

Mary DeBey & Isabelle Barriere

mdebey@brooklyn.cuny.edu & isabelle.barriere@gmail.com

Young children entering our schools at the beginning of the millennium are increasingly inter-connected and interdependent within a global environment. To take full advantage of this interconnectedness, children need to understand people from other cultures and speak languages other than their own. This chapter examines models of dual language programs and classrooms, the effect on early second language learning on literacy and the use of technology in fostering and sustaining dual language programs. Although not a replacement for the traditional activities of childhood, computers in the dual language classroom provide access to cultural and linguistic information that is helpful in dual language learning.

Chapter 5:

Beyond Blackboard: An Update on Course Management Systems

Jeffrey S. Ruth

East Stroudsburg University

jruth@po-box.esu.edu

Web-enhanced, blended, and fully online courses are now a routine component of higher ed offerings. Almost all students in two- and four-year colleges have experienced a course management system (CMS) that organizes and presents the digital face of their course. Blackboard, WebCT, ANGEL, Desire2Learn and other products have dominated this CMS market, but now face real and growing competition from free, open-source programs like Moodle and Sakai. This chapter surveys the terrain of CMS software, describing the logic of the market shift toward open-source products, noting features and pricing differences, and profiling a future e-learning environment in which the CMS – commercial or open-source – may give way to newer formats that have already begun to appear.

Chapter 6:

Best: Practicum for Classrooms as Learning Communities

Fabio Girelli-Carasi

Brooklyn College of the City University of New York

carasi@brooklyn.cuny.edu

Technology as a set of instruments can be used for specific tasks, such as drills, tests, reinforcement and the production of ancillary support materials. Technology as a *medium*, instead, transforms the classroom into a viable learning community. This is possible with a minimal investment of time even when the teacher only knows how to use email and little else. This chapter contains a set of clear instructions for a series of activities whose full responsibility falls on the students, and quickly leads them to increased individual *critical thinking* ability, *collaborative learning* practices, and aspects of a *student-generated curriculum.*

Chapter 7:

The Role of Student Adaptive Self-Regulatory Behavior in the Use of Modern Educational Technology

Brooklyn College of the City University of New York

Manuel Martinez-Pons

mmartinezpons@brooklyn.cuny.edu

Implications for practice of this chapter's findings include the need to include a degree of training in self-regulatory adaptive skill when involving students in the use of educational technology. Much of this instructional resource varies in its effectiveness and cannot always be expected to fully perform its intended function – making it important for the learner to know how self-regulatory adaptive behavior can enable him or her to overcome such limitations when they arise.

Chapter 8:

Technology as Communication Medium for Learning Communities

Fabio Girelli-Carasi

Brooklyn College of the City University of New York

carasi@brooklyn.cuny.edu

To fully exploit the potentiality of the new technologies in the educational setting it is crucial to understand that they are not an amass of discreet tools to be used only for targeted goals. The true strategic advantage of technology is in the fact that it is a *medium* and therefore its holistic effect will be felt in the area of communication within the community that adopts it. If instructors keep their minds focused on this principle, even if they only possess basic technological knowledge they will be able to transform their classes into learning community with minimal effort and maximum results.

References

Cooper, J. D., & Kiger, N. D. (2006). *Literacy: Helping children construct meaning.* (6th ed.). New York: Houghton Mifflin.

Communication From The Commission To The Council, The European Parliament, The European Economic And Social Committee And The Committee Of The Regions i2010- A European Information Society For Growth And Employment (2005). Commission Of The European Communities, Brussels, 1.6.2005 COM(2005) 229 final. Retrieved October 5, 2007, from http://eur-lex.europa.eu/LexUriServ/site/en/com/2005/com2005_0229en01.pdf

Dewey, J. (1916). *Democracy and education.* New York: McMillan.

Goldberg, M. (2006). *Integrating the arts* (3rd ed.). Boston: Pearson.

Intel®Teach to the Future. (2003-9). Retrieved July 20, 2008, from http://www.teachfuture.com/index.aspx

Nation's Report Card – 12th Grade Reading and Mathematics 2005. (2007). IES National Center for Education Statistics, U.S. Department of Education: NCES 2007-4668.

O'Connor-Petruso, S. A., & Lei, J. (in press). History and development of technology in education. In C.L. Lin, D. Ness, & M. Shiegemitsu (Eds.), Section 8 Technologies and global education, *International education and human development: An encyclopedia* (Part I, Section 8, Chapter 1). Armonk: M. E. Sharpe, Inc.

Toward A New Golden Age, How The Internet, The Law And Today's Students Are Revolutionizing Expectations (2004-2005). National Education Technology Plan. U.S. Department of Education. Washington DC: ED Pubs.

UNESCO. (2003). *The use of ICT in education: Trends in Asia and the Pacific.* Retrieved September 8, 2007, from http://www2.unescobkk.org/education/ict/v2/info.asp?id=11012

Visions 2020.2 (n.d.). *Student views on transforming education and training through advanced technologies*. Retrieved June 10, 2007, http://www.nctet.org/Visions2020-2.pdf

Outline:

On the Cover: (left to right)

Letty Vaca, Nazzerine Charles, & Winson Cheong

MSED 2009

City University of New York at Brooklyn College

Cover Design: Genevieve O'Connor Faulk

vievefaulk@gmail.com

Running Head: Embedding Asynchronous and Synchronous Technologies and Sourceware in Course Curricula

Chapter 1:

Embedding Asynchronous and Synchronous Technologies & Sourceware into Curricula

Sharon Anne O'Connor-Petruso

Introduction

In 1954 Lewis M. Terman prophesized that "*within a couple of decades vastly more should be known than we know today about the resources of potential genius, the environmental circumstances that favor its expression, the emotional compulsions that give it dynamic equality ....*" and "*the general awareness of manpower shortage of scientists, engineers ....* " Both Terman's prophesy and concern came to fruition as a dearth of scientists, engineers (especially female), and technologically literate citizens who are cognizant in the hard sciences exist worldwide.

Although we are in our fourth wave of Information and Communication Technologies (ICT) planning, many issues remain unresolved (O'Connor-Petruso & Rosenfeld, 2008) as majority of our nation's youth are neither proficient in Information Age skills nor adept in the hard sciences. Large scale international assessments, such as the Third International Mathematics & Science Study, TIMSS (NCES, 1999) and the Program for International Student Assessment, PISA (NCES, 2008), evince the mediocre scores by American high school students who are well-below the international average –

clearly indicating the need to <u>change</u> how we "teach and learn." Today's instructional methodologies must reflect the critical skills citizens need to be competitive and productive (Dewey, 1916) in the global economy. Hence educators must keep abreast of new and emerging technologies and consistently integrate these tools into curricula.

Educators must also be able to bypass the steadfast issues of outdated infrastructures and limited shelf-life as budgets will always dictate quality of hardware and software in educational environments. A partial solution to this dilemma is for instructors, from elementary to university level, to become adroit in Web 2.0 tools and open-source (free) software – sourceware.

In congruence with both Terman's definition of actualizing talent, or utilizing one's talent in a measurable way, and to help combat the growing "Digital Divide," and /or inability of educational systems to keep pace with emerging technologies, this chapter will focus on the interactive tools of Web 2.0 and sourceware as they have the ability to both inspire youth and provide unlimited opportunities for both students and educators to collaborate on the global platform.

What is Web 2.0?

Web 2.0 is commonly referred to as the second generation of the Internet. It was first coined by web pioneers Dale Dougherty and Tim O'Reilly in 2004 as a result of their brainstorming conference on new and emerging technologies and influx of diversified web sites (O'Reilly, 2009). O'Reilly reported in 2005 that the key objectives of Web 2.0 were the "harnessing of collective intelligence, the sharing of knowledge, and promotion of user contributions" (O'Connor-Petruso, in press).

Although diversified definitions of Web 2.0 still exist today, consensus exists that a key feature of Web 2.0 is its "participatory" attributes as users can directly interact and change web pages due to the creation of Really Simple Syndication/Rich Site Summary (RSS) robust technologies in 1994. These features are categorically evinced in the creation and increasing use of blogs, wikis, and podcasts and are discussed below:

Blog.

A blog is the abbreviated version for web log and is touted for its participatory feature as it instantaneously posts threads in reverse chronological order. A legion of terms has arisen with this new web 2.0 tool such as "blogger, blogging, and blogosphere." A "blogger" is the person who created the blog. "Blogging" refers to posting on the blog. "Blogosphere" refers to the online blogging community at large.

Blog search engine Technorati (n.d.) stated in their 5th annual *State of the Blogosphere 2008 report*, that blogging has become mainstream, citing the 133 million blogs they indexed since 2002 and involved approximately 130+ million bloggers in the past six years alone. However, Technorati also reported for the first time a decrease in blogging. In 2008 there were on average 900,000 blog posts per day, a marked decrease from 1.5 million blog posts per day in 2007.

Blogs are usually created to focus upon a particular topic, subject, and/or interest, and elicit diversified viewpoints. Popular blogs have multimedia features such as text, music, and video (O'Connor-Petruso & Rosenfeld, 2008) as they help prompt discussion (Kelly, 2009). From an educational perspective, whether blogging in blogosphere, course management systems, and/or wikis, the instructor can have a very successful site if he or she consistently monitors the site, sets parameters, and articulates clear behavioral objectives. Similarly, students benefit from instructor maintained blog sites as they have the opportunity to review and discuss assignments as blogs also function as discussion boards.

Free blogging software exists at Wordpress (n.d.) and numerable free commercial hosting sites such as Blogger (n.d.) located on Google and Blog (n.d.) located on Talkspots.Com. One very popular user-friendly blog sourceware is 21Classes Cooperative Learning (n.d.) as instructors are given a classroom index page and a second page for student blogging – both pages are equipped with multimedia attributes.

Wiki

A wiki is a collaborative group web site with various attributes which permit instant web page editing by the members. The wiki webmaster determines whether the web site is private (for select members) or open to the pubic for becoming members. Some of the diversified attributes of today's wiki include multimedia tools, twitters, mathematical equations, and spreadsheets.

Ward Cunningham, an American Programmer, invented WikiWikiWeb (the term *wiki* originates from the Hawaiian language meaning fast and quickly) software in 1994 and posted the first wiki web site in March 2005 (O'Connor-Petruso, in press). Today there are thousands of wiki web sites and they continue to flourish on an international scale. One very popular and burgeoning international wiki is the Wikimedia Foundation, Inc. (2009) web site. It is a non-profit charitable corporation, committed to engaging people around the world to develop and disseminate free information, and is available in nineteen languages. Its online encyclopedia, Wikipedia, is currently the fourth most visited web site in the world (Wikimedia Foundation, 2009). Additional Wikimedia projects include Wikimedia Commons, a repository of media, Wiktionary, a multilingual dictionary, Wikiquote, a repository of international quotations, proverbs, speeches et. al, Wikibooks, free e-books, textbooks, and manuals, Wikisource, a collection of multilingual texts, Wikispecies, a species database, Wikinews, news available in twenty-three languages, Wikiversity, focusing on experiential learning, and MediaWiki, open-source wiki software (Wikimedia Foundation, 2009).

There are several excellent user-friendly commercial wiki sourceware that provide easy page editors, the ability to upload images and files, and add widgets and hyperlinks. One of the most popular commercial wiki's for educators is Wikispaces.com (2009) which hosts over one million wikis with approximately 175,000 in K-12 education. It is extremely user-friendly and both instructors and students enjoy using it. Other popular educational wikis include Wikidot Inc. (n.d.) which is equipped with mathematical equations and Wetpaint.com (n.d.) which provides posting of videos and the ability to create a wetpaint wiki site in less than five minutes.

Podcasts

Podcasts are rich digital media files, either audio or video, and available via web syndication. Podcasts are a direct result of Web 2.0 and RSS technologies. The term podcast is short form for "Personal On Demand broadcast" and is defined in the New Oxford American Dictionary as "a digital recording of a radio broadcast or similar program, made available on the Internet for downloading to a personal player" (Creative Technologies, 2009) - such as the computer or MP3 Player. The motivational and marketing value of incorporating podcasts are clearly recognized by both industry and education as the number of podcasts continues to proliferate. Gonzalez (2005) reported that the number of podcasters will reach 60 million by 2010.

Podcasting has become very popular due to its simplicity and ability to capture digital media files "on demand" with the aid of a "podcatcher" like Apple's iTunes. The Podcaster (person creating and launching the podcast) only needs a standard computer that can play MP3 audio files, recording software or video camera respectively, a file hosting site, and connection to the Internet.

Recording Sourceware & Benefits

Audacity (n.d.) permits recording and editing sounds across both platforms. Precluding the technical expertise gained through digital editing, students are highly motivated to enhance their narratives with music and/or sound effects. Using FreeConference.com (n.d.) and the Internet based telephone system Skype (n.d.), students can record their conference calls and embark upon ongoing communication and endless collaborative projects.

Free Podcast Hosting Site & Benefits

One highly recommended site is PodBean.com (n.d.). In addition to increasing the students aptitude in Web 2.0 tools, the intrinsic and extrinsic value of creating podcasts are evinced in the increased number of student created podcasts. Similarly, on the university level, students value podcasts posted by the instructor as audio and video explanations aid in the understanding of difficult concepts. Instructors can also use podcasts to spur assignments and/or elicit dialogue on discussion board (Kelly, 2009).

Albeit most techies believe we are still in the second generation of the Internet, some technology forecasters believe we have already embarked upon Web 3.0. Mossberg and Swisher (2009) cite the robust synergy of media and technology and offer as evidence Apple's iPhone and iPod Touch and the fact that other leading software companies, like Palm, Research in Motion, Nokia, and Microsoft are *hot* on Apples trail. If these forecasters are incorrect and Web 3.0 has not arrived, the global community will once again place new challenges upon educational leaders and instructors as these new and emerging technologies will require new and advanced skills by students seeking productivity in the global economy.

e-learning

Electronic learning (e-learning) has transformed how we both teach and learn. E-learning should be utilized in every discipline and offers numerable benefits to both the instructor and student. Research concurs (Considine, Horton, & Moorman 2009; Hoic-Bozic, 2009) that when these tools are integrated into the curriculum students become highly motivated and many have increased academic achievement (Nicol, 2009). Similarly instructors report the ease of archiving and updating lessons and documents, offering diverse modes of instructional delivery, and accessing innumerable "free" instructional materials and resources from the Internet.

E-learning communication tools are defined by "type" and "speed of delivery" and are grouped under two broad categories: synchronous and asynchronous. Synchronous delivery refers to the instant transfer of communication, such as two people taking to each other in "real time." Asynchronous delivery refers to "delayed communication," or data that is archived and stored for later access (O'Connor-Petruso, in press).

Synchronous Tools

Common synchronous tools include chat tools and video-based communication tools. Chat tools are versatile, utilized worldwide, and continue to grow in number since the advent of Web 2.0. Some popular free chat tools include Windows Live Messenger (Microsoft, n.d.) and Yahoo Messenger (Softpedia, n.d.). Skype (n.d.) and the Gizmo Project (Free Downloads Center, n.d.) are popular sourceware that allows users to make telephone calls from computer to computer via the Internet using "Voice over Internet Protocol" (VoIP) software.

Video-based communication tools include video streaming, video conferencing, and webcams/web cameras (O'Connor-Petruso, in press). Video streaming or streaming video is compressed data sent in continuous streams over the Internet and displayed to the viewer through a player that decompresses data and sends video to the display and audio to the speakers. Video streaming allows students to interact on a global platform when no videoconferencing technologies exist. There are a plethora of videoconferencing, or video and voiceover IP networks, sites, and directories. Albeit most videoconferencing directories have become proprietary, the A*t&t* (2007) Videoconferencing directory has remained free (since 1995) and includes excellent links to over 47 countries and 1,674 educational sites. Examples of webcams are current traffic and/or weather reports and sports games being viewed in real time. The benefits of integrating video-based communication tools are undisputed as these tools provide knowledge of the world, its peoples, and the opportunity to collaborate globally.

Asynchronous Tools

Common asynchronous tools include electronic mail (e-mail), videos, podcasts, and all the electronic mediums (floppies, compact disc (CD), and flash drive et al.) that store data/content. Lauren Reardon from The Radicati Group, Inc., a technology market research firm, recently stated in her *Email Statistics Report, 2009-2013* that email use continues to mushroom with over 1.4 billion users in 2009, approximately 247 billion messages per day, and projected to increase to 1.9 billion users in 2013 with approximately 507 billion messages per day. Although there are increasingly more videos and podcasts posted online, as seen in the success of YouTube and Teacher Tube, more instructors still need to "catch-up" to the millennial generation and implement these tools into their curriculum.

Where Does Sourceware Come From?

Several organizations, such Red Hat, GNU Project, the Free Software Foundation (Fry, 2008), and the Open Source Initiative (n.d.) are committed to open-source software and provide software, membership, and diversified resources such as tutorials. An excellent interactive tutorial on researching social statistics (Rice, 2009), is provided by JISC, Joint Information Systems Committee, an agency of The Higher and Further Education Funding Councils of England, Scotland and Wales which offers over 60 free online tutorials on a range of subjects.

Mathematics Sourceware

In terms of mathematical freeware, there are a considerable number of sites for statistical software and_resources from both universities and organizations, such as the

Documents Center at the University of Michigan (2008), Independent Evaluators' Webring (2006), and Statcon, Beirwither, & Gabele Software Design (2006). These sites provide in-depth resources, multivariate analysis and graphic packages, and most are available in multiple languages. Free Statistical Software (Corsini, 2009) offers information on learning statistics and statistical news, provides data on several countries, and lists a large number of statistical packages in freeware, limited edition, and open-source. There are also excellent informative sites maintained by professors and social scientists that provide up-to-date statistical software, tutorials, and online learning tools such as Professor Pezullo's Home Page (2005).

Popular content area open-source software includes GeoCebra (n.d.) and Sagemath (2009). GeoCebra sourceware combines geometry, algebra, and calculus, has received several international awards in Europe, and has an active GeoCebraWiki located on their index page. Sagemath.org (2008) is an open-source mathematics interface that combines the power of the proprietary programs of Magma, Maple, Mathematica, and Matlab.

Science Sourceware

StatSci.org, maintained by Gordon Smith (2008), provides a cutting edge directory of statistical packages for analyzing animal abundance. Tools for weather monitoring, simulating sky phenomena, displaying water systems, the Periodic Table, Gemini 10, and earthquakes around the world, among other resources, are located on freeware.intrastar.net (Sawicki, 2009).

Numerable sites for free science software that include diversified resources, downloadable scientific tools, tutorials, and podcasts abound. Recommended sites are the Free Software Directory (n.d.) and the Free Software Foundation (2009).

Web Hosting Sourceware

Precluding free wikis and blog hosting sites, there are very few free commercial web hosting sites as many have become proprietary, like Homestead.com. and Yahoo.com, although they give free trial periods. Fortunately, Google Sites (2009) and Tripod (n.d.) are still offered as sourceware and are very popular due to their user-friendly page editing tools, design templates, and multiple attributes such as email, blogs, and/or widgets.

There are also a few free web hosting services with built in templates that have become very popular with both educators and students. Filamentality (2009) is offered through AT&T Inc. and offers multiple educational activities such as creating hotlists, treasure hunts, and webquests on the subject of your choice. Trackstar (n.d.) originates from the University of Kansas and is touted by educators for it interactive online lessons (termed "tracks") and free database of thousands of tracks by other educators.

Technology Needs Assessment Survey

Due to diversified levels of expertise in technology, ranging from non-existent to highly proficient, and the continuous influx of new and emerging technologies, it is recommended that every instructor, from elementary through university level, give their students a needs assessment survey on Information Age skills (see Appendix) at the onset and end of each semester and/or instructional cycle respectively. The *Technology Needs Assessment Survey* will initially provide the instructor with valuable insight into their students' knowledge and skill level in technology and thus aid the instructor in lesson planning. The second round of administering the survey will evaluate whether or not the instructor's teaching methodologies were effective and increased the students' knowledge and skill level in technology.

The evaluation will ostensibly guide the instructor in designing future behavioral objectives and activities. Due to new and emerging technologies, the process of assessing and evaluating both student knowledge and skills in technology and instructor effectiveness in "turnkeying" these critical skills to their students will remain infinite.

Conclusion

The explosive impact of the chameleonic Internet, Web 2.0 tools, and sourceware has *changed* not only how we teach and learn and motivate our youth but has provided educational leaders with the opportunity to help bridge the Digital Divide as everyone with Internet access can learn Information Age skills and collaborate on the global platform. Albeit the vanguard for implementing these new and emerging technologies initially rests with university professors who create and assess curricula and design teacher education programs – it is "today's teachers" who are charged with preparing "today's youth" to be globally competitive citizens. Thus the main challenge today is for all instructors, elementary through university level, to *catch-up* to the millennials and learn how to use and integrate new and emerging technologies into their curriculum with a focus on sourceware and Web 2.0 tools.

APPENDIX

Technology Needs Assessment Survey

O'Connor-Petruso, S.A.,© 2009

Class:____________________________ Date:_____________________________

Directions: Please fill out Part A: Demographic Survey, Part B: Knowledge & Skill in Technology Survey, and Part C: Rating Scale. Please place your response (choosing the appropriate sub-category number) on the space provided to the right.

Part A: Demographics

1. Gender: (1) Male or (2) Female 1.___

2. Age Range: (1) 8-10 yrs. (2) 11-13 yrs. (3) 18-23 yrs. (4) 24-30 yrs.
(5) 31-37 yrs. (6) 38-44 yrs. (7) 45+ yrs. 2.___

3. Racial Background: (1) Caucasian (2) Hispanic (3) African-American
(4) Asian (5) Other (6) Mixed Heritage 3.___

4. Computers in Home: (1) zero (2) one-two (3) three or more 4.___

5. Computer Use in Home (on average):
(1) Rarely: Once per week.
(2) Infrequently: At least twice per week.
(3) Often: Three-four times per week.
(4) Frequently: At least five times per week.
(5) Daily 5.___

6. My Family Members Use Computer in the Home (on average):
(1) Rarely: Once per week.
(2) Infrequently: At least twice per week.
(3) Often: Three-four times per week.
(4) Frequently: At least five times per week.
(5) Daily 6.___

7. My Instructor Uses Technology in the Classroom (on average):
(1) Rarely: Once or less per semester/cycle.
(2) Infrequently: Two-four times per semester/cycle.
(3) Often: Four-Six times per semester/cycle.
(4) Frequently: Seven or more times per semester/cycle.
(5) Every Class. 7.___

8. My classroom and homework assignments involve using productivity, multi-media applications and Web 2.0 tools (on average):
(1) Zero % (2) 10 % (3) 15-25 %
(4) 26-33.3 % (3) 34-50 % (6) 51-75 %
(7) 76-90% (8) 91-100 % 8.___

Part B: <u>Knowledge & Skill Level in Technology</u>

<u>Directions</u>: Use the key below to rate your skill level and place your response (choosing the appropriate sub-category number) on the space provided to the right.

<u>Key to Skill Level in Technology (on average)</u>:

(1): <u>Nonexistent:</u> Never used/unfamiliar with term/strategy.

(2): <u>Somewhat familiar</u>: No expertise – have tried the tool between 1-5 times per month with limited to no success.

(3): <u>Familiar</u>: Comfortable and successful using the tool – use between 1-5 times per month.

(4): <u>Advanced</u>: Very comfortable and successful using the tool – use between 6-10 times per month.

(5): <u>Expert</u>: Extremely successful – use more than 10 times per month.

<u>Technology Knowledge & Skill Assessment</u>:

1. My knowledge and skill level in word processing is ___. 1.___
2. My knowledge and skill level in creating spreadsheets and graphs is ___. 2.___
3. My knowledge and skill level in creating multimedia (inserting text, sound, animations, and videos) presentations is ___. 3.___
4. My knowledge and skill level in creating graphic organizers (creating charts, graphs, concept mapping etc.) is ___. 4.___
5. My knowledge and skill level in accessing and using streaming videos is ___. 5.___
6. My knowledge and skill level in accessing and using podcasts is___. 6.___
7. My knowledge and skill level in accessing and using wikis is ___. 7.___
8. My knowledge and skill level in creating streaming videos is ___. 8.___
9. My knowledge and skill level in creating podcasts is ___. 9.___
10. My knowledge and skill level in creating web pages is ___. 10.___

<u>Part C: Rating Scale in New & Emerging Technologies</u>

<u>Directions</u>: Use the key below to rate both yours and your instructor's knowledge and skill level in using new and emerging technologies and place your response (choosing the appropriate sub-category number) on the space provided to the right.

<u>Key to Rating Scale (on average)</u>:

(1) <u>Novice</u>: Beginner – use only email and some word processing.

(2) <u>Intermediate</u>: Use email, familiar with productivity and multimedia tools, and interact with web 2.0 tools.

(3) <u>Advanced Intermediate</u>: Very familiar with productivity and multimedia tools, and have created either a blog, wiki, podcast, and/or web page.

(4) <u>Expert</u>: Expertise in productivity and multimedia tools, have created several web 2.0 tools, and is a web master for at least one web site.

1. I rate my knowledge and skill level in new and emerging technologies as ___. 1.___

2. I rate my previous instructor's knowledge and skill level in new and emerging technologies as ___. 2.___

3. I rate my current instructor's knowledge and skill level in new and emerging technologies as ___. 3.___

References

At&t. (2009). *Knowledge Network Explorer*. Retrieved June 1, 2009, from © 2009 Copyright AT&T Intellectual Property Web site: http://www.kn.pacbell.com/index.html

Audacity. (n.d.). Retrieved June 2, 2009, from Sourceforge.net Web site: http://audacity.sourceforge.net/download/

Blogger. (n.d.). Retrieved May 20, 2009, from ©1999-2009 Google Web site: https://www.blogger.com/start?utm_campaign=en&utm_source=en-ha-na-bk&utm_medium=ha&utm_term=blogger&gclid=CKPvqIyt9poCFQOaFQodT1Ud g

Considine, D. Horton, J., & Moorman, G. (2009). Teaching and reading the millennial generation through media literacy. *Journal of Adolescent & Adult Literacy, 52*(6), 471-481.

Corsini, A. (n.d.). *Free statistical software*. Retrieved May 25, 2009, from Copyright © 2001-2009 Andrea Corsini Web site: http://en.freestatistics.info/stat.php

Creative Technologies. (n.d.). *Zencast.* Retrieved June 1, 2009, Creative Technologies Ltd. Web site: http://www.zencast.com/about/

Dewey, J. (1916). *Democracy and education.* New York: Macmillan.

Filamentality. (2009). *Welcome to the knowledge network explorer*. Retrieved June 1, 2009, from © 2009 Copyright AT&T Intellectual Property Web site: http://www.kn.pacbell.com/wired/fil/

FreeConference. (n.d.). Retrieved May 20, 2009, from © 2001- 2009 Global Conference Partners Teleconferencing Web site: http://www.zencast.com/about/

Free Downloads Center. (n.d.). *Gizmo Project 1.1.0.9.0.* Retrieved May 10, 2009, from © 2001- 2009 Free Downloads Center Web site: http://www.freedownloadscenter.com/Network_and_Internet/Voice_Chat_Tools/Gizmo_Project.html

Free Software. (n.d.). Retrieved May 20, 2008, from http://gsociology.icaap.org/methods/soft.html

Free Software Directory. (n.d.). Retrieved April 5, 2009, from the Copyright © 2000-2009 Free Software Foundation Web site: http://directory.fsf.org/category/science/

Fry, S. (2009). *GNU Operating System.* Retrieved April 23, 2009, from GNU Web site: Copyright © 1996-2009 Free Software Foundation Web site: http://www.gnu.org

Gabele, T. (2006). *Freie wissenschaftliche software.* Retrieved May 30, 2008, from © 2006 Tobias Gabele Web site: STAT6http://statistiksoftware.com/free_software.html

GeoCebra. (n.d.). *Welcome to GeoCebra!* Retrieve May 30, 2009, from © GeoCebra 2001-2009 Web site: http://www.geogebra.org/cms/

Gonsalves, A. (2005, July 5). Podcast users expected to reach 60 million in five years. *InformationWeek, InternetWeek.* Retrieved June 1, 2009, from http://www.informationweek.com/news/global-cio/showArticle.jhtml?articleID=165600711

Google Sites. (2009). G*oogle sites overview.* Retrieved May 10, 2009, from © 2009 Google Web site: http://www.google.com/sites/help/intl/en/overview.html

Hoic-Bozic, N., Mornar, V., & Boticki, I. (2009). A blended learning approach to course design and implementation. *IEEE Transactions on Education, 52*(1), 19-30. Abstract retrieved June 8, 2009, from EBSCO database.

Kelly, R. (2009, February 27). Using video clips to stimulated discussion in online courses. *Faculty Focus*. Retrieved March 25, 2009, from http://www.facultyfocus.com/articles/asynchronous-learning-and-trends/using-video-clips-to-stimulate-discussion-in-online-courses

Microsoft. (n.d.). *Windows Live Messenger 8.1*. Retrieved May 15, 2009, from © 2009 Microsoft Corporation Web site: http://www.microsoft.com/downloads/details.aspx?familyid=d78f2ff1-79ea-4066-8ba0-ddbed94864fc&displaylang=en

Moodle. (2008). *Welcome to the moodle community!* Retrieved January 20, 2009, from http://moodle.org/

Mossberg, W., & Swisher, K. (2009, May 26). Welcome to web 3.0. *D All Things Digital*. Retrieved June 1, 2009, from The Wall Street Journal.Digital Network Web site: http://allthingsd.com/

National Center for Education Statistics. (2008). *Highlights from PISA 2006: Performance of U.S. 15-year-old students in science and mathematics.* (NCES 2008-016). Washington DC: U. S. Government Printing Office.

National Center for Education Statistics. (1999). *TIMSS: Overview and key findings across grade levels.* (NCES 1999-01). Washington DC: U. S. Government Printing Office.

Nicol, D. (2009). Assessment for learner self-regulation: Enhancing achievement in the first year using learning technologies. *Assessment & Evaluation in Higher Education, 34*(3), 335-352.

O'Connor-Petruso, S. A. (in press). Electronic/synchronous & asynchronous learning. In C. Lin, D. Ness, & M. Shigemitsu (Eds.), *International education and human development: An encyclopedia* (Part I, Section 8, Chapter 3). Armonk: M. E. Sharpe, Inc.

O'Connor-Petruso, S. A., & Rosenfeld, B. (2008). Effective strategies for integrating technology and the tools of web 2.0 when limited by budget, infrastructure, and shelf life. *31st Proceedings of the Association for Educational Communications and Technology, Vol.2, On the Practice of Educational Communications and Technology*, FL. Retrieved May 25, 2009, from http://www.aect.org/publications/proceedings/2008.asp?id=2

Open Source. (n.d.). *Open source initiative*. Retrieved May 20, 2009, from opensource.org Web site: http://www.opensource.org/

O'Reilly, T. (2005). *What is web 2.0. Design patterns and business models for the next generation of software.* Retrieved May 10, 2009, from © 2009, O'Reilly Media, Inc. Web site: http://oreilly.com/pub/a/oreilly/tim/news/2005/09/30/what-is-web-20.html?page=1

Pezzullo, J. (2009). *Web pages that performs statistical calculations (StatPages.net*). Retrieved June 5, from John C. Pezzullo's Home page Web site: http://statpages.org/index.html

PodBean.Com. (n.d.). Retrieved May 10, 2009, from © 2006-2009 Podbeam.com Web site: http://www.podbean.com/start-audio-blog?sourceid=goog_audio_blog

Reardon, R. (2009, May 6). *Email statistics report, 2009-2013*. Retrieved May 20, 2009, from http://www.radicati.com/?p=3237

Rice, R. (2000-2009). *Welcome to internet for social sciences.* Retrieved May 12, 2009, from intute: virtual training suite Copyright 2000-2009 Web site: http://www.vts.intute.ac.uk/he/tutorial/social-statistics

Sage. (2009). *Open source mathematics software*. Retrieved June 7, 2009, from http://www.sagemath.org/

Sawicki, A. (2009). *Science – Free Software*. Retrieved April 30, 2009, from © 1997-2009 Alan Sawiki Web site: http://freeware.intrastar.net/science.htm

Skype. (n.d.). Retrieved May 10, 2009, from © 2009 Skype Limited Web site: http://register.skype.com/en-us-skype?cm_mmc=google/latsearch-_-AMERC-and US|ENG|RELAT-_-CONBD-_-kwid=JFT047477|creative=3020038364&gclid=CK6c1ozh6ZoCFQuU7QodyCWNBg

Softpedia. (n.d.). *Yahoo! Messenger 9.0.02162*. Retrieved May 10, 2009, from © 2001-2009 Softpedia Web site: http://www.softpedia.com/get/Internet/Chat/Instant-Messaging/Yahoo-Messenger.shtml

State of the Blogosphere/2008. (n.d.). Retrieved May 20, 2009, from © Technorati, Inc. Web site: http://technorati.com/blogging/state-of-the-blogosphere/

Statistical Packages: Free. (2009). Retrieved March 23, 2009, from Copyright © Gordon Smyth 1996-2007 Web site: http://www.statsci.org/free.html

Statistical Resources on the Web (n.d.). Retrieved July 12, 2008, from the University of Michigan MLibrary Documents Center Web site:

http://www.lib.umich.edu/govdocs/stats.html

Talkspot.Com. (n.d.). Retrieved May 11, 2009, from © 2009 Talkspot Blogging Web site: http://www.talkspotblogs.com/aspx/m/0

Terman, L. M. (1954). Discovery and encouragement of exceptional talent. *American Psychologist, 9*(6), 221-230.

Top Science Software. (2009). Retrieved May 1, 2009, from the dmoz open directory project Web site: http://www.dmoz.org/Science/Software/

Trackstar. (n.d.). *4teachers*. Retrieved May 18, 2009, from Copyright. © 2000-2008 ALTEC at the University of Kansas Web site: http://trackstar.4teachers.org/trackstar/index.jsp

Tripod. (n.d.). Retrieved May 30, 2009, from © 2009 Locus Inc. Web site: http://www.tripod.lycos.com/

Wikimedia. (2009, January 9). *Frequently asked questions.* Retrieved May 22, 2009, from Wikimedia Foundation, Inc. Web site: http://wikimediafoundation.org/wiki/About_Wikimedia

Wikimedia. (2009, April 19). *Home.* Retrieved May 22, 2009, from Wikimedia Foundation, Inc. Web site: http://wikimediafoundation.org/wiki/Wikimedia:About

Wetpaint. (n.d.). Retrieved May 22, 2009, from wetpaint.com, inc.© 2009 Web site: http://www.wetpaint.com/

Wikidot.com. (n.d.). *Wikidot.* Retrieved May 25, 2009, from 2007-2009 Copyright Wikidot Inc. Web site: http://www.wikidot.com/

Wikispaces. (2009). *Wikis for everyone*. Retrieved February 20, 2009, from Tangient LLC Web site: http://www.wikispaces.com/

Wordpress.Com. (n.d.). Retrieved January 23, 2009, from WordPress.org Web site: http://en.wordpress.com/

21Classes Cooperative Learning. (n.d.). Retrieved June 1, 2009, from © 2004-2009 21Publish Inc. Web site: http://www.21classes.com/

Chapter 2:

Instructional Strategies for Integrating Technology into the Curriculum

Barbara Rosenfeld

Introduction

One of the great challenges that teachers face is how to present knowledge in a way that students will easily comprehend. If students were more like robots and less like humans, this task would be significantly simplified! However, each student brings a unique background to the classroom and each learns in his/her own manner. Teachers can present instruction in a variety of ways; indeed, teachers have many factors to consider when preparing a lesson. It is hoped that a teacher's educational background will help him/her to make informed decisions based on knowledge of the field.

Just as there are several approaches to instruction, there are many ways to integrate technology tools into the curriculum. Paramount to whichever way is chosen is the golden rule of technology use: whatever and however technology is used, **it must be aligned with the lesson objectives**. If the technology does not support the learning objective, its use should be excluded. Technology that is used merely because it is fun or because the students like it is not in line with good educational practice. There needs to be a purposeful rationale for classroom activities.

Technology tools can be used in any part of a lesson to help learners learn. For example, technology can help: to motivate/interest students in the topic, to aid in introducing new material, to present the focal point of the instruction, to review the topic, to enhance or enrich the curriculum for students who need more challenging work, or to provide review and/or practice for students who would profit from additional time with a topic.

Technology tools help students to reach their goals. For example, technology can provide access to resources (i.e., pertinent web sites and online library resources), aid in the writing process (i.e., word processing programs), improve presentations (i.e., presentation software), facilitate data collection, organization, and interpretation (i.e., spreadsheet programs), and assist in collecting graphic images, oral histories, and interviews, (i.e., still and video cameras, audio recordings, cell phones). Students who make use of journals, logs, blogs, wikis and other such web resources become part of a collaborative group that can help each other in the learning process. As students write and reflect on their learning, they can share their thoughts and reflect on the thoughts of others. Technology tools can also help the teacher to provide more exciting and interesting lessons with the use of interactive whiteboards, videotapes, and presentation software. Teachers may also record information/lessons as podcasts that students can access at any time.

When a teacher plans a lesson with technology, it is beneficial to have some guidelines. Here are some guidelines to follow, based on Roblyer's (2003; 2006) model for technology integration as well as the author's experience from teaching this topic.

1. Determine the goals and objectives of the unit/lesson.
2. Determine the focus of the student learning to be addressed and the relative advantage of doing so with technology tools.
3. Choose appropriate assessment strategies.

4. Select appropriate instructional approach and integration strategies.

5. Resolve the logistics of preparing the environment and instructional materials.

6. Evaluate and revise the strategies used.

This chapter will address each of these guidelines.

1. Goals and Objectives - Determine the Goals and Objectives of the Unit/Lesson

Think about the goals and objectives for the unit/lesson. How can the objectives best be accomplished? How can technology aid or enhance these objectives? Remember the golden rule of technology use: the technology must be in alignment with the unit/lesson goals and objectives. Further, goals and objectives must be aligned with the stated curriculum standards. There may be national, state and local standards as well as standards in specific disciplines.

For example, in the area of mathematics, the National Council of Teachers of Mathematics (NCTM) has developed principles and standards for the teaching of mathematics in grades K-12 (see the NCTM web site at http://nctm.org/). States have also developed mathematics standards for their students. Particular localities may have another set of standards. Teachers should be familiar with all the pertinent standards for their particular situations and keep them in mind when they are designing their lessons.

The International Society for Technology in Education (ISTE) has developed national technology standards and performance indicators for students (ISTE, 2007) and national educational technology standards for teachers (ISTE, 2008). Following are the six student standards that should guide technology use in the classroom:

1. creativity and innovation
2. communication and collaboration

3. research and information fluency
4. critical thinking, problem solving, and decision making
5. digital citizenship
6. technology operations and concepts

There are five standards that mandate teachers to:

1. facilitate and inspire student learning and creativity
2. design and develop digital-age learning experiences and assessments
3. model digital-age work and learning
4. promote and model digital citizenship and responsibility
5. engage in professional growth and leadership

When technology tools are used to aid teaching, these standards should be applicable. (For more detail visit the ISTE web sites listed in the reference list.)

Teachers should be careful to state objectives in terms of observable behavior so that the outcomes can be assessed. Objectives should include verbs like "describe," "demonstrate," "write," "compare," etc. Words like "understand," "appreciate," and "know" should be avoided because they have an internal focus and are difficult to measure (Roblyer, 2006).

2. Learning Focus and Relative Advantage – Determine the Focus of the Student Learning to be Addressed and the Relative Advantage of Doing So with Technology Tools

Will the use of technology make the learning more effective or more efficient? Will it promote higher level thinking and problem solving? Will students find the learning more engaging? If the use of technology does not aid the learning process (and/or is not aligned with the learning objectives), it is inappropriate to use.

Web resources can help the learner to perceive relationships that would be difficult to see without the use of the technology. For example, students who are beginning to learn about geometry concepts such as parallelograms, rectangles, and triangles can go to the National Council for the Teaching of Mathematics web site called "Illuminations" by clicking on: http://standards.nctm.org/document/eexamples/chap5/5.3/index.htm#applet. There they can find activities that will aid in building their conceptual understanding of the relationships among polygons. They can explore a dynamic parallelogram to see if it can be manipulated into a rectangle or a square and experiment with a dynamic rectangle to see if it can be manipulated into a parallelogram. This would be difficult to do with regular manipulatives, but it is easily accomplished on the computer. The web site can help students make predictions and conjectures and then let students prove them. This makes the learning less abstract and more concrete and can facilitate student understanding of the properties of various geometric shapes.

Another way that web resources can be useful is by providing students with practice and feedback. Sometimes students find paperwork to be boring or tedious whereas practicing in an interactive computer environment with immediate feedback might be more motivating, especially if it is in a gaming or challenging format (i.e., How many you can do correctly?). For example, a sample mathematics objective might be for the elementary student to measure a particular length with a metric ruler. This is a practical skill, not a cognitive task. A teacher could pass out rulers and have students practice measuring different linear objects. Depending on student ability, all students might be successful in performing this task. If the teacher has many students who need help, it might be difficult to provide assistance to all. Students who need further practice with measurement could visit this web site: http://www.funbrain.com/measure/index.html (Funbrain.com, n.d.). The web site provides three levels of difficulty, depending on the skill level required. Encouraging feedback (i.e., Nice Work! Right!) is given with correct answers, and feedback with the correct answer is given with incorrect answers. Students can work on this task independently (or in small groups), freeing the teacher to walk around and give support and assistance as needed.

What is the advantage of using a web site like this? Much depends on whether a teacher has the time to check each student's work and provide feedback. If the teacher is able to check each student's work, this web site might not be so useful. The activity does not provide much in the way of creativity and higher thinking skills, but it could aid a student in learning how to measure correctly, a necessary life skill. Students will still need some hands-on practice to insure that they know how to place a ruler properly when they measure. However, students often find that online work is very motivating and a welcome change of pace from the challenges of the everyday class (or paper and pencil tasks that students perceive as boring). The teacher has to decide whether this technology tool would be useful toward reaching the lesson objectives.

Thirty years ago, if students were asked to report on current events, they sometimes read a single article in a newspaper. The article was clipped and brought to school for presentation and discussion. Students often had the same article. Today if students are asked to report on current events, there are abundant web resources that can be used (in addition to the newspaper that might be found in the home or library). Students could check a single newspaper web site (i.e., The New York Times at http://www.nytimes.com/) to get some basic information. Students might also want to compare the reporting of two or more newspapers (online) to see how each one handled the material. Television, radio, and YouTube web sites might also be checked (i.e., www.cnn.com) along with blogs and wikis on the topic. Students could report not only on the content, but also on the way the information is presented. Is the information unbiased? Was the reporter at the scene or dependent on second-hand information? What did others have to say about the topic on a blog? (Who are these "others" and what is *their* expertise?) A richer discussion can arise from looking at more than one resource. This may help to broaden students' perspectives about issues. Without the use of technology, students' resources would be more limited in scope.

Students can also use the technology to produce a project. For example, students might be asked to discuss a character in a book and create a hypothetical scene to portray this character in a new way, perhaps changing the time or the place. Students in the past could do this by simply writing a new scene that might be submitted to the teacher or acted out in class. Today they could act out the scenario and tape it with a videocamera that they could show to their classmates. They could also post the video on YouTube and share it with the world. When students know that others will look at and critique their work, they tend to be very engaged with it (Leister, 2009).

3. Assessment - Choose Appropriate Assessment Strategies

When planning instruction, a teacher can make use of an instructional systems design model, a way of systematically designing the instruction. In most teacher-designed instruction the teacher plans instruction and then, after the instruction has been given, makes up a test based on what was "covered" during the unit. With the instructional systems design model, after the goals and objectives have been determined, but before the instructional strategies have been developed or presented, the assessment is created.

The assessment should match the objectives of the lesson; otherwise there is no way of knowing whether the objectives have been met! According to Gagné (Aronson & Briggs, 1983) there are five varieties of learning, each requiring a different type of instruction (and a matching objective). These learning types include intellectual skills (i.e., using concepts to create new rules), motor skills (i.e., ability to move in a particular way), verbal information (stating information), cognitive strategies (i.e., ability to find new solutions to problems), and attitudes (i.e., choosing to behave in a particular way). The assessment should be in line with the learning type and should be observable. An example of an assessment for a motor skill might be: The student will be able to throw a basketball into the hoop with 80% accuracy. An example for an intellectual skill might be: The student will demonstrate how to find the average of 3 numbers. If the objective is a motor skill, it should not be assessed with a multiple choice test, but rather with a physical test, for example, to determine how well the student can throw the basketball. If the objective is to demonstrate knowledge, the student must be given the opportunity to do so, either in writing or orally.

It should be noted that there is much standardized testing in schools today, and these tests are often graded by a machine. Many tests are of the short answer variety (multiple choice, true-false, matching) simply because they are easier to grade (especially by machine) than other types of assessments. Teachers know that essays require more time and effort (and more knowledge of English grammar and spelling) to grade than short answer tests. If the learning objectives are more factual, short answer tests may be appropriate. However, if problem solving and higher level learning skills are being assessed, short answer tests may not be the best way to evaluate student learning. In these cases, essays, portfolios and presentations may be more appropriate. Students often use technology to prepare these kinds of assessments; checklists and rubrics are often used to assist in these tasks (Roblyer, 2003; 2006).

Rubrics

Rubrics provide a scale to grade students for each of the various parts of their learning. Points are awarded based on the student's level of achievement. Depending on the objectives, rubrics may look at behavioral and socialization skills as well as content areas. The teacher can decide how many points each element of the learning is worth.

It should be noted that socialization skills (i.e., How well does the student work with others? How does the student contribute to the group?) are important because they are needed in the workplace. More people lose jobs because of a lack of social skills than for performing poorly on the job (ERIC Clearinghouse on Handicapped and Gifted Children Reston VA, 1987; Houghton & Proscio, 2001; McLellan, 2008). Schools need to provide collaborative experiences to prepare students for future job opportunities in the workplace.

Portfolios

Portfolios are collections of student work that represent performance over time (Smaldino, Lowther, & Russell, 2008). Portfolios may include exemplary work, they may include a variety of pieces to show growth to the current level of expertise, or they may be a combination of both. Students are usually asked to reflect on their learning and include these reflections (sometimes in logs, blogs, journals, etc.) in their portfolios. This helps to give the students greater insight into their own learning (Newby, Stepich, Lehman, & Russell, 2000; 2006). Portfolios are often rated with performance checklists, attitude scales, product-rating checklists, and rubrics (Smaldino et al., 2008).

Electronic portfolios (e-portfolios) allow the student to store work in many formats (i.e., audio, video, graphics, and text.) E-portfolios are an optional way to simplify the storage, display, and organization of student artifacts, making the collection less cumbersome and retrieval less complicated. Online portfolio programs are often available for a fee.

4. Instructional Approach and Integration Strategies – Select Appropriate Instructional Approach and Integration Strategies.

After the teacher has decided what the goals, objectives, and assessments will be, s/he must decide which instructional strategies and which technology tools will best help the students to reach these objectives. Instructional strategies may be categorized as teacher-directed or student-centered (Smaldino et al., 2008). There are two main teacher-directed approaches – one based on behaviorism and the other based on an information-processing (cognitive) approach. In both cases, the teacher provides information that students "learn" followed by practice and an assessment.

What is learning? According to Driscoll (2005), learning is "…a persisting change in performance or performance potential that results from experience and interaction with the world" (p. 1). We can define the term; we know that a change takes place in a particular way. However, when we ask when and how this change takes place; that is, how learning occurs, we find that these questions are not easily answered. Educators have pondered and continue to ponder these questions because there are no clear-cut answers.

We still await the genius who can figure out how we learn so that we will know how to best present instruction. Perhaps in the future there will be a machine that will transmit information into the brain so there will be no need for learning and instruction. At present there are no definitive answers, but there are hypothetical constructs based on intuition, investigation, and research to help explain what is yet unknown.

Teacher-Directed Learning

Behaviorists believe that learning occurs when one can observe a change in an individual's behavior. Behaviorists are not concerned with the reason for a particular behavior, in fact they feel that the rationale for the behavior is unnecessary information and may even be problematic (Driscoll, 2005). For example, when high school students study for a test, are they studying to learn the material and gain more knowledge, or are they studying to get a good grade so they can get into an ivy league college? Behaviorists do not care what the reason is; what matters is how well they perform on the test. The outcome is what is important because that is the way one can gauge whether learning has taken place.

Whereas the behaviorists focus on the external environment – where learning is revealed as a behavior that can be observed – the cognitivists focus on what transpires internally in one's brain. Cognitivists believe that learning occurs when the mind is "actively engaged in processing information…." (Smaldino et al., 2008, p. 41.) Cognitivists believe that the mind organizes information in some way to be retrieved when needed later. The computer is often used as a model to describe this approach to learning. When information is received, the computer has to store it away, just as a person has to store information in his/her mind. When the information is needed, the computer has to bring it out of storage so that it can be called into service, just as a person does. Cognitivists explore how humans actually select and store information, how a person can take something new and relate it to his/her past knowledge, and how humans retrieve this information from their memory so they can apply it to new situations (Roblyer, 2003; 2006).

Student-Centered Learning

Constructivism is a third learning perspective. Constructivists believe that students learn by constructing knowledge from participation in realistic experiences. One of the characteristics of a constructivist lesson is that it includes problems that students find relevant and interesting and that often have more than one solution, mimicking the complexity of problems in the real world. It should be noted that more time and more diverse skills and knowledge are often required to solve these problems than with teacher-directed learning. Usually students work in collaborative groups, much as they would in a workplace environment, thereby preparing students for life in the "real world." Each student brings with him/her a unique background of varying experiences, helping to solve the problem with multiple approaches (Newby et al., 2000; 2006).

In student-centered learning approaches the students are responsible for directing their own learning with guidance from the teacher. The teacher acts as a facilitator rather than the giver of knowledge. The teacher's role changes from being the "sage on the stage" to one of "guide on the side." As students work in groups the teacher can circulate through the classroom offering assistance as needed.

Integrating Technology

Technology can support any learning approach. Again, teachers must examine the goals and objectives to determine what the student outcomes will be and then find appropriate technology to support those outcomes. The teacher needs to think about technology needs. For example, will more than one computer be needed? Where will the lesson take place? The teacher also needs to be aware of student abilities and behaviors. For example, how much guidance will the students require? Will students be able to work collaboratively? Resolving these questions leads to the next guideline.

5. Logistics – Resolve the Logistics of Preparing the Environment and Instructional Materials

Classroom Environment

A classroom may have a single computer, a small bank of computers, or there may be access to enough computers so that each student is able to work on his/her own machine. If a single computer is used, the teacher may use a projection unit so that all students will be able to view the screen image at the same time. Some software programs were designed for the one-computer classroom (such as the *Decisions, Decisions* software from Tom Snyder Productions). Alternately, students could take turns at the computer so that all are able to cycle through to have a hands-on experience. Some classrooms are now equipped with interactive white boards that allow the user to easily manipulate web sites, prepare and make presentations.

If there is a small bank of computers, teams of students could work on them at one time, cycling through so that all students have a turn for a hands-on experience. The computers can be configured to accommodate an activity. For example, each computer could be prepared as a different work station (i.e., one with a software program, another with an Internet site, etc.).

Many schools have a computer lab (or a library media center) with multiple banks of computers so that every student can work at a separate station. The configuration of these labs is often in rows to accommodate as many computers in the space as is possible. However, this arrangement is not usually optimal for discussion purposes since students cannot see one another and must twist and turn to face each other. The teacher must account for this obstacle to discussion when preparing a lesson.

Where will the lesson take place? Will it be in the classroom, in a computer lab, or in the library media center? The classroom is generally the easiest place to work, as students do not have to travel to another space and prior arrangements may be unnecessary. Sometimes a portable laptop computer cart can be brought into the classroom. Handheld technology with probes is sometimes used in the field to gather data for later evaluation (i.e., to test water quality). The teacher needs to be familiar with the process for acquiring technology tools in the school (i.e., Whom to contact to secure the technology tools for the time they are needed.). However, if available technology tools are insufficient for the classroom, another space could be used. If a computer lab or library media center is required, the teacher needs to know what arrangements have to be made to secure the lab (or library media center) at the needed time. The teacher also should be aware of what software the school provides for classroom use.

When technology is mentioned for classroom use most people think of computers, but there are other technology tools that help student learning such as still or video cameras. Depending on the nature of the activity, students could work with cameras in the classroom, in the field, or in a studio. Interactive white boards, mentioned earlier, can be portable or permanently mounted in the classroom. They make it easy to manipulate web sites, make presentations, and show relationships in mathematics and other subject areas. Audio/videotapes (CDs, DVDs, etc.) are additional tools that can help in the learning process. Again, the teacher needs to know what equipment is available, how to use it and how to access it. If a teacher does not have this information, s/he should ask administrators for it.

Student Grouping

Teachers must also decide how to group students for an optimal learning situation. Will students work individually on tasks? Will they work in a small group? How will students be grouped and what is the rationale for doing so? For example, students might be grouped in twos because there are only 15 computers for a class of 28 students. The teacher might want to pair students who are academically the same so that both students can contribute equally to the task. Another option is to pair those who are academically different so a stronger student can help a weaker one. Students might be grouped in homogeneous or heterogeneous groups of three or four, where each student has a particular task (cooperative learning). Will the groups be mixed with regard to gender? Are there students who do not work well together or some that work "too well" together and socialize more than they work? Teachers must think about all of these things when preparing a lesson.

There should be a reason for grouping students in a particular way. For example, sometimes students work in a lab where there is one computer per student, yet the teacher asks that students work in small groups. Why? Perhaps the teacher is interested in having students collaborate on their work to practice socialization skills. Perhaps the teacher wants each student to make a particular contribution to a group project. Research has shown that students prefer to work with a partner when they are learning something new (Rosenfeld, 1992). There are many reasons for having students work together. Sometimes the group will work on one computer; at other times each student will work on his/her own computer and come together for discussion or to compose a final report or presentation. There are many choices, but the teacher has to be aware of not only the goals and objectives, but these other factors as well so that s/he may select the best options for his/her particular learners.

6. Evaluate and Revise – Evaluate and Revise the Strategies Used

After a lesson has been taught, the teacher should reflect on it to determine if the goals and objectives were met and whether there might be a better way to approach the topic. What worked well? What didn't work as well? Students can also be surveyed (informally or formally) to gauge their perceptions about the lesson. If the teacher is pleased with the outcomes of the lesson, s/he may want to use the lesson again in the future. If the teacher feels that there is room for improvement, then s/he should think about ways to tweak the lesson to make it better. (Sometimes "tweaking" is insufficient – a major overhaul is necessary.)

There are times when a teacher wants to address a problem in a classroom to try to improve it in a more formal way. A teacher may decide to systematically improve some area in which students are having difficulty through an action research project. This may be accomplished with the use of technology tools such as video or audio recording to gather data (Roblyer, 2003).

Phases of Technology Integration

It should be noted that the guidelines for technology integration assume that the teacher already knows how to use the technology tools. A teacher is probably not going to use tools with which s/he is unfamiliar. Teachers who are recent graduates are more likely to be savvy technology users than those who graduated a decade or more ago. Technology integration usually does not happen because a teacher wills it, although intent is an important factor. It takes time and experience for a teacher to figure out how best to use technology tools in the classroom. It is helpful if the teacher has seen technology tools used effectively in a classroom so that s/he is aware of what good technology integration should be. It is therefore appropriate for colleges and universities to model good classroom technology use, especially for their future teachers. An excellent web site may be found at InTime (1999-2001), the result of a PT3 (Preparing Tomorrow's Teachers to Use Technology) Grant at the University of Northern Iowa. One can watch videos of exemplary K-12 teachers modeling technology use with their classes and one can observe interviews with the teachers as they discuss the lessons. It is an award-winning resource.

Leister (2009) suggests that there are three phases of educational technology use, dependent in part on teacher expertise and availability of equipment. In Phase I (Using Technology to Present Dynamic Lessons) teachers use technology to supplement lessons by presenting them dynamically, using PowerPoint and other multimedia software to engage their learners. Generally these lessons are teacher-directed, allowing the teacher to be in control while gaining experience with the technology tools.

When teachers are competent and confident users themselves, they may ask students to use the technology tools to access information and perhaps to work on projects that require an end product. Phases 2 and 3 are more student-centered. In Phase 2 (Using Technology to Access Information) students use online sources to access content knowledge. They may use web resources, work with software to improve skills or perhaps try a web-based treasure hunt. These are "read-only" activities; there is no creative product as a result of this activity and the access is to mostly Web 1.0 resources.

In Phase 3 (Using Technology to Produce and Share Products) students produce a product to be shared. This often requires that students use technology to access information (Phase 2) before they are able to complete their task. The product is some kind of publication to a wider audience than the class, such as an entry on a blog, a web-based product, or a movie. Students must create a product, and they often communicate and collaborate with others to accomplish this. Phase 3 is actually an on-going cycle: The student is a producer, then a publisher, and then the student becomes part of an audience and peer-reviewer for the products of others. Students are contributing not only to their own knowledge base, but they are contributing to the knowledge base of others.

Concluding Thoughts

Schools should provide their teachers with ongoing training and support in the use of classroom technology tools. One of the huge barriers to technology use is a lack of support in the schools. If a teacher is having difficulty using the technology or if the technology does not function as it should, technical support should be easily available. If teachers face too many challenges with technology use, the relative advantage disappears.

If it is too much bother, the computers (and other technology) become dust collectors – in some classrooms they are broken dust collectors. Administrators who believe that technology tools are necessary in the classroom will do their best to hire technical support personnel and educational technology specialists who can provide the know-how to fix the equipment and provide the professional development to train teachers.

In a perfect world, the technology will always work exactly as planned. However, in our imperfect world the technology sometimes fails to function properly. If no one can quickly correct the problem, it is important that the teacher have a backup plan so that the entire lesson is not lost. What can a teacher do if the Internet links are malfunctioning? Backup transparencies can be made for an overhead projector. Teachers can prepare and distribute handouts of important pages and/or PowerPoint slides can be made. There is a software program (WebWhacker 5.0, 2008) that enables the teacher to take pictures of web pages so that they can be displayed later on a computer without the need for an Internet connection. This is also effective for sharing useful web sites that might be inadvertently blocked by filtering systems and it is a way to build a virtual web site library for later classroom use.

At the end of the 20th century schools began to use what is now termed Web 1.0 technology tools to help students learn. At the beginning of the 21st century, Web 2.0 technology has made its way into the classroom. As new technologies become available, educators need to think of ways to harness this technology to engage and enable students to learn more efficiently and effectively.

References

Aronson, D. T., & Briggs, L. J. (1983). Contributions of Gagné and Briggs to a prescriptive model of instruction. In C. M. Reigeluth (Ed.), *Instructional-design theories and models: An overview of their current status* (pp. 75-100). Hillsdale, NJ: Lawrence Erlbaum Associates.

Driscoll, M. P. (2005). *Psychology of learning for instruction* (3rd ed.). New York: Pearson Allyn & Bacon.

ERIC Clearinghouse on Handicapped and Gifted Children, Reston, VA. (1987). *Developing social vocational skills in handicapped individuals.* Retrieved January 7, 2008, from http://www.ericdigests.org/pre-927/social.htm. (ERIC Document Reproduction Service No. ED287263)

Funbrain. (n.d.). *Measure it!* Retrieved March 5, 2009, from © 2000-2009 Pearson Education, Inc. Web site: http://www.funbrain.com/measure/index.html

Houghton, T., & Proscio, T. (2001). Hard work on soft skills: Creating a "culture of work" in workforce development. (Report). Philadelphia, PA: Public/Private Ventures. (ERIC Document Reproduction Service No. ED461737)

ISTE. (2007). *The ISTE National Educational Technology Standards (NETS•S) and Performance Indicators for Students.* Retrieved December 2, 2008, from http://www.iste.org/Content/NavigationMenu/NETS/ForStudents/2007Standards/NETS_for_Students_2007_Standards.pdf

ISTE. (2008). *The ISTE National Educational Technology Standards (NETS•T) and Performance Indicators for Teachers.* Retrieved December 2, 2008, from http://www.iste.org/Content/NavigationMenu/NETS/ForTeachers/2008Standards/NETS_T_Standards_Final.pdf

Leister, J. (2009). *Three phases of educational technology*. Retrieved February 28, 2009, from http://www.sanjuan.edu/webpages/jleister/

McLellan, W. (2008, October 30). *Job seekers lack basic social skills*. Retrieved January 7, 2009, from http://www.montrealgazette.com/jobs/job-listings/seekers+lack+basic+social+skills/920033/story.html

National Council for Teaching of Mathematics. (2004). *E-example 5.3: Exploring properties of rectangles and parallelograms.* Retrieved January 7, 2009, from http://standards.nctm.org/document/eexamples/chap5/5.3/index.htm#applet

New York State Education Department. (2008). *Welcome to mathematics, science, technology.* Retrieved July 25, 2008, from New York State Education Web site: http://www.emsc.nysed.gov/ciai/mst.html

Newby, T. J., Stepich, D. A., Lehman, J. D., & Russell, J. D. (2000). *Instructional technology for teaching and learning: Designing instruction, integrating computers, and using media* (2nd ed.). Upper Saddle River, NJ: Merrill Prentice Hall.

Newby, T. J., Stepich, D. A., Lehman, J. D., & Russell, J. D. (2006). *Instructional technology for teaching and learning* (3rd ed.). Upper Saddle River, NJ: Pearson Merrill Prentice Hall.

Roblyer, M. D. (2003). *Integrating educational technology into teaching* (3rd ed.). Upper Saddle River, NJ: Merrill Prentice Hall.

Roblyer, M. D. (2006). *Integrating educational technology into teaching* (4th ed.). Upper Saddle River, NJ: Pearson Merrill Prentice Hall.

Rosenfeld, B. (in press). Technology and the curriculum. In C.L. Lin, D. Ness, & M. Shiegemitsu (Eds.) Section 8 Technologies and global education, *International education and human development: An encyclopedia.* Armonk, NY: M.E. Sharpe.

Rosenfeld, B. H. (1992). The effects of individualized and small group learning approaches on the motivation of college students. (Doctoral dissertation, University of Missouri – Columbia). *Dissertation Abstracts International, 54*(7), 2491a.

Smaldino, S. E., Lowther, D. L., & Russell, J. D. (2008). *Instructional technology and media for learning* (9th ed.). Upper Saddle River, NJ: Merrill Prentice Hall.

Resources

CNN.com. (2009). Cable News Network, Turner Broadcasting System, Inc. Retrieved March 3, 2009, from www.cnn.com

Dockterman, D. (2008). Decisions, Decisions (Version 5.0) [Computer software]. Watertown, MA: Tom Snyder Productions. Retrieved March 4, 2009, from http://www.tomsnyder.com/products/Product.asp?SKU=DECDEC

Funbrain.com – Measurement game. (2000-2009). Pearson Education, Inc. Retrieved March 3, 2009, from http://www.funbrain.com/measure/index.html

InTime. (1999-2001). Retrieved March 3, 2009, from http://www.intime.uni.edu/

The New York Times. (2009). Retrieved March 3, 2009, from http://www.nytimes.com/

WebWhacker 5.0. (2008). Retrieved February 28, 2009, from http://www.bluesquirrel.com/products/webwhacker/

YouTube. (2009). YouTube, LLC. Retrieved March 3, 2009, from http://www.youtube.com/

Chapter 3:

Literacy – A Critical Constituent for Successful Globalization

Meral Kaya, Sharon A. O'Connor-Petruso, & Fabio Girelli-Carasi

Introduction

It is essential to be literate in the Global Age and technology promotes literacy education. In order to meet the needs of the global citizen, teachers need to be able integrate digital tools to enhance and promote their teaching and learning of literacy in the classroom. The necessity of teaching literacy is apparent as English is a universal language. Technology will support the teaching curriculum, which will lead to successful globalization.

Teaching literacy carries the responsibility of raising a literate "GLOBAL" citizen. However the conceptions of what it means *to be literate* have changed rapidly in response to the demands of globalization. Literacy used to be defined as "reading and writing" but has now expanded to include listening and speaking, competencies in mathematics and technology, and proficiency in visual, communication, and calculation skills (Cooper & Kiger, 2006; Goldberg, 2006).

The broad definition of literacy also brings changes in literacy teaching methods, approaches, and materials. "In order to understand and construct meaning from any print or communication media, it is necessary to have language skills and multiple sign systems and symbols represented in technology" (Velmont, 2003, p.23). Literacy teaching should provide these skills, systems, and symbols. Hence technology needs to be integrated into literacy teaching - specifically in reading workshops, writing workshops, guided reading, and assessment. In order to meet the needs of the global citizen, teachers need to be able to integrate digital tools to enhance and promote learning. Therefore, literacy teachers need to be trained on how to utilize technology for teaching and learning. This chapter will explore the tools of technology that enhance literacy teaching and learning.

Technology and Literacy Activities

The communication tools of technology have profound ramifications on what it means to be literate in the Global Age as well as literacy instruction, approaches, techniques, hands-on activities, and assessment. As Watts-Taffe & Gwinn (2007) claim, technology brings new changes within itself, which as a result changes "the nature of what it means to be literate as well as demanding a new and more important roles for the teachers" (p. 3). Today, educators and professionals of staff development are constantly searching for answers on how they will utilize technology, integrate these tools into their curriculum, and hopefully turnkey these digital skills and strategies to their students.

One of the tools of technology in literacy instruction is software. Englert, Zhao, Collings, & Romig (2005) conducted a study that explored the effectiveness of internet-based software on the improvement of reading performance. They utilized TELE-Web software, Technology Enhanced Learning Environment on the Web, to teach sight words to a group of at-risk first grade students. Students were exposed to high frequency words in cloze test format and target words were presented in short passages. At the end of the study students successfully recognized the sight words. Thus this study shows that internet-based software can play an important role in reading programs as it can be used to teach as well as assess reading and vocabulary.

Another popular literacy activity is WebQuests. WebQuests necessitate students' active involvement with inquiry through internet resources (Dodge, 1995; Ikpeze & Boyd, 2007; Wang & Hannafin, 2008). WebQuests provide opportunities for students to research, think critically, gather data, and analyze information. Through well-designed structure such as introduction, task, information resources, process, guidance, conclusion (Dodge, 1995), WebQuests serve as an effective and functional internet resource that teachers can utilize to encourage their students to be critical thinkers as well as researchers. We, as teacher trainers, should be aware of WebQuests importance and be able to integrate it into our teacher education programs. Wang and Hannafin (2008) provide guidance for scaffolding pre-service teachers in integrating technology, such as WebQuests, into curricula.

Technology and Children's Literature

Literacy learning requires "meaning making, language study and inquiry-based learning" (Kasten, Kristo, McClure & Garthwait, 2005, p. 11) and children's books serve as great sources to accomplish this process. The goal of reading programs should include not only teaching children to learn to read but to also raise lifelong readers who enjoy reading (Kiefer, Hepler & Hickman, 2009) which can be accomplished by assigning quality stories. Teachers also use these stories to teach language arts. Due to the Digital Age, language arts instruction now comprises viewing and visually representing skills along with speaking, listening, reading, and writing skills. Educators must also try to convey meaning from non-print media such as web pages, movies, and CD ROMs. Language arts instruction has morphed to include all the skills that help children become literate.

Children's literature books have also extended their boundaries to digital life. In addition to the numerable databases dedicated to children's literature, electronic books or e-books, and various applications pertaining to the teaching of language arts, teachers and teacher candidates are shown strategies on how to use computer-based animation to practice oral language as well as vocabulary teaching, and how to implement blogs, wikis, and discussion boards to hone their reading and writing skills and post fiction and nonfiction stories.

Reading stories from print is not the only way to engage with literacy activities. Stories on screen provide opportunities of literacy activities especially for young readers (specifically the millennial generation) who are already accustomed to the digital world. Levy (2009) studied 12 young children to assess their screen story engagement and experience. Her research shows that these young children developed both the skills to read the text as well as comprehend the meaning. Levy suggests that schools include multimedia as part of their reading programs as children's exposure to technology-based stories will decrease when print literacy replaces on screen story engagements.

Today technological tools are viewed as powerful ways for having children excel in reading (Ediger, 2004). One strategy is to have children listen to the CD ROM version of a book before reading it from print. This especially helps English Language Learners (ELLs) as well as serving as "a role model for oral reading in terms of stress, pitch and pronunciation for the student, positive examples for structure of the English as well as for syntactical models" (p. 379).

Language arts instruction is now presented and delivered through technology. Cole and Williamson (2005) used electronic journals, e-journals, as a reading writing activity to enhance fourth grade students' comprehension. Cole, who teaches literacy methods courses, worked with Williamson, a fourth grade teacher in another state. Together they created an e-journal partnership between the college and elementary school to explore if e-journals had a positive effect on fourth graders' comprehension skills. Their focus was on questioning, connecting, visualizing, and summarizing as well as the college students' understanding of fourth graders' reading and writing skill development. They used children's literature books, specifically books by Gary Paulson, for their e-journals. Their results showed that both college and elementary school students benefited from the e-journal experience. This research demonstrated that technology can be a powerful tool for literacy instruction and learning.

Children's literature also serves as a great source to learn words and build vocabulary knowledge (Roe & Ross, 2006). The textual and visual contexts provided by children's literature help students make predictions on the meaning of words as well as remember and retain the related meaning. Labbo, Love, & Ryan (2007) found that learning words from incidental and direct exposure to vocabulary through computer related activities were especially helpful to children who either came from low socio-economic backgrounds or had learning problems. The Labbo et al. (2007) study entitled "Vocabulary Flood" focused on a "five day instructional cycle that uses computer technologies and related activities" (p. 582) to help students learn and retain meaning and usage of new words. Two methods are used in this study: (I) Digital Language Experience Approach (D-LEA developed by Labbo, Eakle & Montero, 2002 which is based on the Language Approach, by Allen, 1999) which consists of four steps: (1) setting-up the experience, (2) photographing the experience, (3) composing a multimedia photo essay or story and, (4) engaging in follow up activities, and (II) Using digital whiteboards that serve as an interactive touch screen/monitor when connected to a computer (Solvie, 2003). Students are provided opportunities to "notice, and make meaningful words they encounter during storybook read-alouds by thinking about them, talking about them, and using them in writing on multiple occasions" (p. 583). Recycling (using) words through retelling, taking digital photos of activities and performing in the activities related to photos help students not only learn words but also retain the information as well as "use the words in multiple occasions" (p. 583). The Vocabulary Flood study shows that technology related activities provide effective learning, better comprehension, and the ability to use new words in various contexts.

Analytical learning also occurs in the instructional strategy of literature study. Literature discussion is a common practice in language art instruction. "Literature study brings students together for an in-depth discussion of a work of fiction and nonfiction" (Fountas & Pinnell, 2001, p. 252). Through literature discussion students make connections and respond to many aspects such as characters, setting, genre, author's style, and many others. This powerful technique can also be implemented through technology. Moreillon & Tatarchuk (2003) conducted an action research project on online literature discussions between undergraduate students studying children's literature from geographically distant classrooms. They chose literature to create discussion between these schools through the online support tool *Polis* (similar to Blackboard) developed by Sally Jackson. Although researchers initiated the discussion they let students continue without any interference. The result of the study showed active engagement in discussion as well as learning. The researchers suggest that "shared virtual classrooms create a powerful communication and learning environment for promoting multicultural understanding and technology enabled literacy" (p. 7).

Technology Literacy and Assessment

Assessment is a salient part of instruction. Assessment and instruction go hand-in-hand in the learning process. Assessment is defined as "the process of gathering information about a student's performance" (Cooper & Kiger, 2008, p. 457). Cooper and Kiger (2008) discuss two kinds of assessments: summative, "pertaining to a final assessment" (p. 497) and formative, "pertaining to ongoing assessment" (p. 501).

There are formal and informal assessments as well. The tools that are used in formal assessment include norm-referenced criterion (standardized) tests. Informal assessment contains various assessment tools such as observation, surveys, checklists, running records, portfolios, and all the reading and writing activities provided in the class. Although standardized tests are important in that they provide "snapshots of students' learning" (Nettles, 2006, p. 96), informal assessment gives the teacher the opportunity to look deeper into the student's progress and thus provide better reflection and feedback for future instruction. In the process of informal assessment, it is critical that students are actively involved in the learning process and participate in the review of their assessments.

In the 21st century teachers must search for various instructional tools and methods that will provide students with opportunities to be "active learners" as well as fit their learning styles. The constructivist perspective views students as "active learners" when they can make meaning of the given information (Woolfolk, 1998).

How does technology assist the authentic assessment process? Teachers and students and work on web based reading and writing activities (Cooper & Kiger, 2008) individually or in groups and can communicate via email, discussion boards, blogs, wikis, and/or podcasts. Students can also create their own web sites showcasing their work where they have the opportunity to hone their assignments and respond to peers, while instructors monitor their interaction and assess their work.

Technology also helps in standardized test preparation. Computer programs for these tests allow obtaining results faster and help educators and teachers see how their students, classrooms, districts, and states perform. Teachers can also assess the effectiveness of the literacy programs through comparisons of content specific variables.

Technology also provides opportunities for teachers to store, retrieve, and obtain various recorded information about students performance and progress and to make better instructional decisions. Artifacts, anecdotal/observational notes, interview and conference notes can be recorded through technology and easily retrieved by the educator. Watts-Taffe et al. (2007) discusses the role of technology, as "monitoring" (p. 89) students' progress from one reading/writing stage to the next level and keeping track of what has been read and who has been reading.

The most popular technology-based assessment tool is digital portfolios. Portfolios are authentic and extremely functional in monitoring the students' progress in learning. Digital portfolios help both teachers and students monitor and assess their progress over time. Digital portfolios also provide self-assessment for both teachers and students (Rickards, Diez, Ehley, Guilbault, Loacker, Reisetter, Hart, & Smith, 2008). According to Hicks, Russo, Autrey, Gardner, Kabodian & Edington (2007) "digital portfolios can foster critical, creative and reflective teaching for teachers" (p. 450). They claim that digital portfolios should not be viewed as an end product, but more of a "reflection and growth" opportunity (p. 451). Technology has thus made portfolios more accessible for both teachers and students as it is available via the web. Students also have the ability to add, change, organize, browse, retrieve, and examine/analyze their portfolios.

Conclusion

Teaching literacy carries the responsibility of raising a literate "GLOBAL" citizen. This means our students must come through our educational system competent in the universal language of English, in the hard sciences, and adept in technology, visual, and communication skills. Educators are to be made aware of the importance of technology and the necessity of incorporating it into curricula. Teacher trainers and educators are responsible for informing their students on current techniques, applications, and Web 2.0 tools. Thus educators must be cognizant of the new changes and approaches in teaching literacy and the new tools of technology that motivate and enhance literacy teaching and learning.

References

ALAN Online. (n.d.). Retrieved June 10, 2009, from © 2006–2007 ALAN Online Web site: http://www.alan-ya.org/

Allen, R. V. (1999). Using language experience in beginning reading: How a language experience program works. In O. Nelson & W. Linek (Eds.), *Practical Classroom Applications of Language Experience* (pp. 41-47). Boston: Allyn & Bacon.

American Library Association. (n.d.). Retrieved June, 2009, from © 2009 American Library Association Web site: http://www.ala.org/

Britannica. (n.d.). Retrieved June 10, 2009, from © 2009 Encyclopedia Britannica, Inc. Web site: http://www.britannica.com/

Celebrating Cultural Diversity through Children's Literature. (n.d.). Retrieved June 10, 2009, from http://www.multiculturalchildrenslit.com/

Cole, J. E., & Williamson, D. (2005). E-Journals: Bridging the literacy partnership. *Illinois Reading Council, 33*(2), 32-43.

Cooper, J. D., & Kiger, N. D. (2006). *Literacy: Helping children construct meaning.* 6th edition, Boston: Houghton and Mifflin.

Cooper, J. D., & Kiger, N. D. (2008). *Literacy assessment: Helping teachers plan instruction.* 3rd edition. Boston: Houghton and Mifflin.

Database of Award-winning Children's Literature. (2009). Retrieved June 10, 2009, from Copyright© 1997-2009 Lisa R. Bartle Web site: http://www.dawcl.com/

Dodge, B. (1995). Some thoughts about WebQuests. *The Distance Educator, 1*(3), 12-15.

e-Buddies. (n.d.). Retrieved June 12, 2009, from © Best Buddies International Web site: http://www.ebuddies.org/

ED.gov. (n.d.). Retrieved June 8, 2009, from US Department of Education Web site: http://www.ed.gov/index.jhtml

Ediger, M. (2004). Challenge in children's literature. *College Student Journal, 38*(3), 379-379.

Englert, C. S., Zhao, Y., Collings, N., & Romig, N. (2005). Learning to read words: The effects of internet based software on the improvement of reading performance. *Remedial and Special Education, 26*(6), 357-371.

ePals Exchange Projects. (n.d.). Retrieved June 8, 2009, from © 2009 ePals, Inc. Web site: http://www.epals.com/

Fountas, I. C. & Pinnell, G. S. (2001). *Guiding readers and writers.* Portsmouth, NH: Heinemann.

Goldberg, M. (2006). *Integrating the arts* (3rd ed.). Boston: Pearson.

Hicks, T., Russo, A., Autrey, T., Gardner, R., Kabodian, A., & Edington, K. (2007). *Journal of Adolescent and Adult Literacy, 50*(6), 450-458.

Ikpeze, C., & Boyd, F. (2007). Web-based inquiry learning: Facilitating thoughtful literacy with WebQuests. *The Reading Teacher, 60*, 644-654.

ISTE. (2009). Retrieved June 8, 2009, from Copyright © 2009 International Society for Technology in Education (ISTE) Web site: http://www.iste.org/

Kasten, W. C., Kristo, J. V., McClure, A. A., & Garthwait, A. (2005). *Living literature: Using children's literature to support reading and language arts.* Columbus, OH: Pearson.

Kids' Space. (n.d.). Retrieved June 8, 2009, from CopyRight 1995-2008 Web site: http://www.kids-space.org/

Kiefer B. Z., Hepler, S., & Hickman, J. (2007). *Charlotte Huck's Children's Literature* (9th ed.). Boston: McGraw Hill.

Labbo, L. D., Love, M. S., & Ryan, T. (2007). A vocabulary flood: Making words "sticky" with computer response activities. *The Reading Teacher, 60*(6), 582-588.

Labbo, L. D., Eakle, A. J., & Montero, M. K. (2002). Digital Language Experience Approach: Using digital photographs and software as a Language Experience Approach Innovation. *Reading Online, 5*(8). Retrieved June 8, 2009, from http://www.readingonline.org/electronic/elec_index.asp?HREF=labbo2/index.html

Levy, R. (2009). 'You have to understand words…but not read them': young children becoming readers in a digital age. *Journal of Research in Reading, 32(1),* 75-91.

Literacy Matters. (2009). Retrieved May 22, 2009, from © 2003-2008 Education Development Center, Inc. Web site: http://www.literacymatters.org/

Literature Circle Resource Center. (n.d.). Retrieved May 28, 2009, from © 2006 Katherine L. Schlick Noe Web site: http://www.litcircles.org/

Moreillon, J., & Tatarchuk, B. (2003).Digital discussions: La esperanza in the shared virtual classroom. *Reading Online, June*, 7-22.

NCAM. (n.d.). Retrieved June 10, 2009, from http://ncam.wgbh.org/

Nettles, D. H. (2006). *Comprehensive literacy instruction in today's classrooms.* Boston: Pearson.

Read Write Think. (June 10, 2009). Retrieved June 10, 2009, from Copyright 2002-2009, IRA/NCTE Web site: http://www.readwritethink.org/

Rickards, W. H., Diez, M. E., Ehley, K., Guilbault, L. F., Loacker, G., Reisetter-Hart, J., & Smith, P. C. (2008). *The Journal of General Education, 57*(1), 31-50.

Roe, B. D., & Ross, E. P. (2006). *Integrating language arts through literature and thematic units.* Boston: Pearson.

Solvie, P.A (2003). The digital whiteboard: A tool in early literacy instruction. The *Reading Teacher, 57*, 484-487.

The Bulletin. (n.d.). Retrieved May 29, 2009, from Copyright 2009, Board of Trustees of the University of Illinois Web site: http://bccb.lis.illinois.edu/

The Children's Literature Web Guide. (n.d.). Retrieved June 8, 2009, from Copyright © 1994-2001 David K. Brown Web site: http://www.acs.ucalgary.ca/~dkbrown/

Valmont, W. J. (2003). *Technology for literacy teaching and learning.* Boston, MA: Houghton and Mifflin.

Wang, F., & Hannafin, M. J. (2008). Integrating WebQuests in preservice teacher education. *Educational Media International, 45*(1), 59-73.

Watts Taffe, S., & Gwinn, C. B. (2007). *Integrating literacy and technology: Effective practice for grades K-6.* Guilford Publications, New York, NY.

Wikipedia. (n.d.). Retrieved June 10, 2009, from Wikimedia Foundation, Inc. Web site: http://en.wikipedia.org/wiki/Main_Page

Woolfolk, A. E. (1998). *Educational Psychology*. Boston: Allyn & Bacon.

Select Web Sources Children's Literature Authors' Websites

You can find information and various activities relating to their children's books on the following authors' websites:

Author's Name	Web Site	Reference
Avi	www.avi-writer.com	Avi. (n.d). Retrieved May 29, 2009, from © 2002-2009 Avi Web site: http://www.avi-writer.com/

Judy Blume	www.judyblume.com	Welcome to my Web site. (n.d.). Retrieved May 28, 2009, from © 2007 Judy Bloom Web site: http://www.judyblume.com/
Jan Brett	www.janbrett.com	Welcome to Jan Brett's Home Page. (n.d.). Retrieved May 30, 2009, from © Jan Brett 1996-2009 Web site: http://www.janbrett.com/
Eric Carle	www.eric-carle.com	Welcome to the Official Eric Carle Web Site. (n.d). Retrieved May 29, 2009, from http://www.eric-carle.com/
Tomie DePaola	www.tomie.com	Tomie.com. (n.d.). Retrieved May 30, 2009, copyright © Tomie dePaola Web site: http://www.tomie.com/
Mem Fox	www.memfox.com/welcome.html	Mem Fox. (n.d.). Retrieved June 10, 2009, from http://www.memfox.com/welcome.html
Gail Gibbons	www.gailgibbons.com	Gail Gibbons. (2009). Retrieved May 28, 2009, from 1998-2009©Gail Gibbons Web site: http://www.gailgibbons.com/
Gary Paulsen	www.randomhouse.com/features/garypaulsen/	Gary Paulsen. (n.d.). Retrieved May 28, 2009, from Copyright © 2009, Random House, Inc. Web site: http://www.randomhouse.com/features/garypaulsen/
Patricia Polacco	www.patriciapolacco.com	Patricia Polacco.com. (n.d.). Retrieved June 10, 2009, from http://www.patriciapolacco.com/
Audrey Wood	www.audreywood.com/mac_site/clubhouse/clubhouse_page/clubhouse.htm	The Audrey Wood Club House. (n.d.). Retrieved June 10, 2009, from http://www.audreywood.com/mac_site/clubhouse/clubhouse_page/clubhouse.htm

Web Sources on Literacy and Children's Literature

International Society for Technology in Education

This site is for K-12 teachers and administrators to explore ways of integrating technology into students' learning.
http://www.iste.org

U.S Department of Education: Teachers Site

This site provides students, teachers, parents and administrators various teaching and learning resources including technology.
http://www.ed.gov/teachers/landing.jhtml

ALAN Online: Assembly of Literature for Adolescents

This is an independent assembly of NCTE, that serves teachers, authors, librarians, publishers, teacher educators, students, and others who contribute to young adult literature.
http://www.alan-ya.org

Kids Space

This is a great site where children worldwide explore and communicate through using new technologies.
www.kids-space.org

Literature Circle Resource Center

This site provides literature circle related information, projects, ideas, extension activities as well as professional resources about literature circles.
www.litcircles.org

Read Write Think

This site provides practices and recourses in reading and language art instruction.
http://www.readwritethink.org/

Literacy Matters

This is a great site for teachers and parents who need to help struggling readers and writers. It provides various resources and activities as well as articles.
http://www.literacymatters.org/

The Bulletin of the Center for Children's Books

This site provides children's books, review journals including a Center for Children's Books.
http://bccb.lis.illinois.edu/

National Center for Accessible Media

This is a media access site for people with disabilities. It provides audio story books on CDs to help learners with reading difficulties.
http://ncam.wgbh.org/

Database of Award-Winning Children's Literature

This database helps students, teachers, and librarians access a reading list of quality children's books plus an award winning book list.
http://www.dawcl.com

American Library Association

This site contains resources and information on children's and young adult literature for teachers, students, and librarians.
http://www.ala.org

The Children's Literature Web Guide

This site provides internet resources related to books for children and young adults.
http://www.acs.ucalgary.ca/~dkbrown/

Multicultural Children's Literature

This is a great site that provides an access to the list of multicultural books from distinct cultural groups.
http://www.multiculturalchildrenslit.com

ePals Exchange Projects
This site helps students, teachers, and classrooms connect and collaborate globally through various projects.
http://www.epals.com/

e-Buddies: A Best Buddies International Program
This is an email pen pal program that connects via email and pairs an intellectually challenged with a peer who does not have intellectual disability.
http://www.ebuddies.org

Online Encyclopedias:
Britannica: http://www.britannica.com/
Wikipedia: http://en.wikipedia.org/wiki/Main_Page

Chapter 4

Globalization, Technology And The Education Of Young Children: Understanding The World In Two Languages

Mary DeBey & Isabelle Barriere

Brooklyn College of the City University of New York

Introduction

Early Childhood teachers structure environments in which young children develop strong healthy bodies, deep understandings, a strong sense of self and the ability to live among and communicate with others. Through art, stories, movement, dramatic play, games, sand, teacher led activities and blocks, children make meaning of the world in which they live. Within these rich program environments, children in the United States, as they do in countries around the world, are best served in our global society when they are given the opportunity to understand the world in more than one language.

Children entering early childhood programs are increasingly inter-connected and interdependent within a global environment. They have accessibility to technology which greatly influences their access to information, how they play and work and communicate with others in the neighborhood and far away places. To take full advantage of this interconnectedness, children need to understand people from other cultures and the languages they speak.

This chapter examines models of dual language programs and classrooms, the effect on early second language learning on literacy and the use of technology in fostering and sustaining dual language programs. Although the National Association for the Education of Young Children (NAEYC) defines early childhood education as programs working with children from birth to age eight, the focus of this chapter is on infant, toddler and pre-school programs.

Dual Language Learners

Dual language learners are children who are learning two languages at the same time or a second language while continuing to develop their first. For many decades, second language learning for young children in the United States focused on teaching English to children who did not speak English or had limited skills. As the number of immigrants increased the number of programs such as English as a Second Language (ESL), bilingual, Limited English Proficient (LEP), English language learners (ELL) and children who speak a language other than English (LOTE) increased.

More recently, dual language programs for young children have centered on maintaining the home language of children with a home language other than English. This is most evident in the Improving Head Start for School Readiness Act of 2007 (P.L. 110-134) which places emphasis on supporting each child's home language and improving outreach and increased enrollment and quality of services for children and families in communities with speakers of languages other than English (LOTE). Given the positive research supporting the learning of more than one language in early childhood, the third and present wave is to afford all children in the United States, including those in which English is the home language, the advantage of knowing a second language. As all children in the United States are guaranteed a Kindergarten experience, the learning of a second language is seen as the birth right of all children in the United States.

Young children learn language in either a simultaneous or sequential manner. Simultaneous acquisition of two or more languages is defined as the learning of different languages at the same time, beginning at birth. Sequential acquisition to a second language is beginning a second language at or after three years of age. When learning languages later in life, children learn language differently. Once a child is three, the sequential language learners have learned conceptual knowledge in their first language which they use in learning the second language (Ervin-Tripp, 1974). Children learning a second language later in life, translate the meaning of words into the first language.

Dual Languages in Early Childhood Classrooms

Early childhood education instruction as well as curriculum differs from that of older children in distinct and significant ways. In later schooling, content is broken into discreet subject areas (i.e., reading, geometry, biology) and taught at different and discreet times during the day or week. In early childhood classrooms, the curriculum is integrated with the most efficient learning happening during a free choice time where materials and activities are carefully considered for their developmental value before being placed in the classroom. The amount of extended free choice time expands as children require more time to finish ever increasingly complex buildings, paintings, and stories. In quality early childhood programs, the curriculum, materials and instructions are developmentally appropriate and culturally and linguistically responsive. The teachers are responsive to the individual needs of children and differentiates not only instruction, but materials (Copple & Bredekamp, 2009).

The difference in instruction and curriculum in the education of young children has important implications for second language learning. The second language is not taught as a subject. Rather it is a means to communicate as both languages are incorporated into the daily schedule. During free choice and teacher-directed activities, meal times and meetings, young children are given the opportunity to understand and communicate in more than one language.

Dual language programs have different configurations. The ideal early childhood dual language program has a 50/50 mix of children speaking each language resulting in a large number of peers fluent in each language (Curtain & Dahlberg, 2003). In programs with more than one teacher in the room, each teacher may stay in a given language throughout the day and children hear and use each language depending on the teacher with whom they are interacting (DeBey & Bombard, 2007). Some programs alternate the languages on a daily basis while in early elementary schools, where separate subject areas begin to surface, the languages are often subject specific (i.e., math may be in Spanish and science taught in English).

While learning a second language is often arduous in later elementary and secondary school, young children learn languages best through play. Play is important for children's social, emotional, and cognitive development, as well as a reflection of their development (Copple & Bredekamp, 2009). Using their knowledge of child development, early childhood teachers are informed by their understanding of child development and provide materials and activities that allow children to hear and speak in two languages.

The concern within the early childhood community that the emphasis on academic performance has led to programs that are more teacher-directed and didactic with little time for play (Elkind, 2008) has implications for dual language learning. In the adult world, play may be considered the antithesis of work, but for young children, it is the way they work and therefore learn. Play is how children learn most efficiently. Piaget (1962), Vygotsky (1966), Bruner (1983), and Webb (2009) each understood the importance of unstructured play in the physical, social, emotional and cognitive development of young children. Early childhood classrooms where children have time to play with peers, especially peers with a language other than their own, are the most ideal for learning a second language.

As in all early childhood programs which support and foster language development, dual language teachers scaffold experiences and language so that children can communicate and be successful. The teacher presents work and words at a level that children can make the reach to the next level. Scaffolding along with techniques such as gestures, speaking more slowly, the use of simpler words and gradually increasing the complexity allow children to more easily learn a second language.

Languages and Culture

The commitment to a second language is crucial, which language (i.e. Mandarin, Spanish, Urdu) is less so. Children do not become proficient in a language when a program exposes children to many cultures and languages for short periods of time. As with learning the first language, children need consistent and extended time in the language to gain proficiency (DeBey et al., 2007; Genesee, Paradis, & Crago 2004). In dual language early childhood classrooms, children need ample opportunities to hear and practice language; building vocabulary and concepts as they go through their day.

In choosing a language, one must consider the number of children in the class speaking the language, the presence of the language and culture in the community, the availability of teachers speaking the language, the availability of materials and the support of parents. Choosing a language is more difficult if there are several home languages in the classroom or if all of the children in the classroom speak English.

If there are several home languages in the classroom a choice must be made. Jessner (1999) found that learning a second language more readily prepares you for learning a third. Therefore children with a home language other than the target language (second language) of the program, has the benefit of learning a third language. As with all children, the home language of children learning English and a third language needs to be consciously supported. In programs where few or no children speak the target language, teachers need to be aware that children revert to their common first language in play, slowing down the language learning process. During play, teachers can model the target language to encourage the peer to peer interaction.

Successful early childhood second language programs are additive, meaning a second language is added to the first, rather that subtractive in which the first language is reduced in order to teach the second language. Second language programs which are additive are beneficial for young children (Winsler, Diaz, Espinosa & Rodriguez, 1999).

Language and culture are intrinsically linked and therefore the teaching of a language can not be separated from the culture. The language not only reflects the culture but cultural experiences shape how we see the world. Evidence of culture can be apparent (i.e., dress, music, food) but is also more subtle and expressed in the use of words, rhythms, gestures, patterns of speech and silence (Sanchez, 1999). Teachers who understand and reflect the culture assist the language learning process. Teachers can use technology to provide cultural experiences and an understanding of nuances of culture that may not be readily available within the program or community. Museums around the world have opened up collections to which teachers, parents and children have access. For example, the Smithsonian in Washington D.C. (www.smithsonianeducation.org), the Metropolitan Museum or Art in New York City (metmuseum.org) and the Asian Art Museum in San Francisco (www.asianart.org/index.html) are available.

Dual Language Learning for All Children

Espinosa (2008) found that all young children are capable of learning two languages and that becoming bilingual has long-term cognitive, academic, social, cultural, and economic benefits. Knowing two languages provides greater mental flexibility and assists in acquiring math concepts and solve word problems (Zebsko & Antunez, 2000). Although teachers and parents may be concerned that children learning two languages may get confused or fall behind in English, children who learn two languages simultaneously have language milestones similar to children learning only English (Stechuk, Burns, & Yandian, 2006; Petito, et al., 2001). As infants and toddlers, the vocabulary in each language may be smaller but the total number of words across both languages is comparable to that of English-only-speaking children (U.S. Dept. of Health & Human Services/ACF, 2001).

Teachers of young children growing up in low income English-only families often feel the need to concentrate on English. They are concerned that the learning of a second language will be a distraction from the learning of content in English for children who come to school with far few vocabulary words than children from middle and upper class families. The lack of vocabulary words does have a negative affect on reading and writing, however, a second language increases knowledge about language and therefore has a positive affect on the first language. To stop children from low-income families from falling further below their peers, they must have the same opportunity to learn a second language. As early childhood programs provide a language-rich environment in multiple languages, a child with fewer vocabulary words has the opportunity to hear and use two languages.

Early childhood programs that are play based are especially beneficial for children from low income families (Miller, 1983). Miller found that when children with similar IQ's were placed in one of three types of early childhood programs: a Montessori program, a traditional play-based program and a didactic (academic skill and rote learning based) program, the didactic preschool programs were less effective than the traditional and Montessori preschool programs in promoting academic achievement. The difference remained evident in eighth grade math and reading achievement (Miller, 1983).

There is also concern that children with disabilities will be confused if exposed to more than one language. Candelaria-Greene (1996) found that children with disabilities can learn a second language and function as well in both languages as their peers who do not have disabilities.

Dual Language Learning and Early Literacy

Language and literacy are interconnected. This is true when children are learning one or two languages. The language that surrounds the play in early childhood classrooms, including vocabulary and listening comprehension leads to higher literacy skills (Dickinson & Tabors 2001). Teachers foster language and literacy as they "feed in" language when young children play with puzzles or build in the block corner. Cognitive and conceptual abilities increase as children develop language (Nelson, 1996) and in turn the higher level conceptual knowledge requires an expanded lexicon. Language and literacy development also affects social development. For example, Snow, Burns, & Griffin (1998) found that a child's early social development predicts long-range outcomes regarding literacy.

Quality early childhood dual language classrooms are literacy rich classrooms. An environment filled with books and stories, play and language, writing and art materials and time to explore and communicate with adults and peers supports literacy. The ability to use and hear language within the classroom is an important part of language learning. Christie and Roskos (2006) found that play in early childhood classrooms is directly linked to the development of important literacy skills such as oral language where there is a connections between socio-dramatic play and oral language development, phonological awareness in which there is an awareness of the sounds of language, print awareness in which areas of play have rich reading and writing opportunities, and content knowledge which is inherent in play and is supported by a teacher's involvement in the play.

Second Language Development Sequence

When children learn their first language, they understand words (receptive language) before they say words (expressive language) (NICHD 2000). For example a baby will look at a bottle upon hearing the word before he says the word bottle. In a dual language classroom, children follow the same sequence. Therefore because a child is not yet speaking the second language, it does not mean that a child is not beginning to understand the target language. Tabors & Snow (1994) propose a developmental sequence for young children learning a second language:

1. First-language use. Children continue to use their first language although they are in the second-language situation. At this time they are gathering awareness that there is a language with which they are not familiar.
2. Observational and listening period. During this time children show awareness of the second language but may not try it. Receptive language is evident at this time.
3. Telegraphic and formulaic speech. Children use individual words and phrases in the target language. Telegraphic speech is the use of a few content words to mean an entire sentence. Formulaic speech is the use of a set phrase they have heard in a like situation.
4. Productive language use. Children use the second language in an increasingly complex fashion.

As the sequence suggests, second language learning, as it does with the first language, takes time. With the emphasis on literacy in elementary schools today, second language learning is often seen as a luxury or a hindrance to the improvement of test scores. Without this pressure in early childhood programs, children can be afforded this time.

Using Technology to Support Second Language Learning

Young children are fascinated with computers. Young children imitate their parents and older siblings use of them in play just as they have imitated talking on the telephone in the dramatic play area for decades. Children do more than play computers however. There are computers designed specifically for children as well as computer software. Computer programs, designed specifically for young children, are most often described as educational. The educational value varies, with most programs teaching concrete knowledge such as vocabulary, numbers and letters which is quite easily taught without the use of computers. However in second language learning a teacher can choose software that specifically addresses a vocabulary and sentence structure need for a particular child or a small group. There is an abundance of software in many target languages (i.e. Spanish, French, Mandarin) developed to teach and support a second language. Software designed for children often contains a suggested age range, however parents and teachers must evaluate the developmental appropriateness for an individual child and remember that software is no substitute for learning languages through play, rather a tool that is integrated into a rich dual language program.

With sound teacher's judgment and expertise, computers can enhance second language learning and literacy. The principles of developmentally appropriate practices and culturally and linguistically sensitive curriculum should be used in deciding how, when and how often computers are used (NAEYC, 2008). The computer is most effective in early childhood classroom when they are "integrated into the regular learning environment and used as one of many options to support children's learning" (NAEYC, 2008).

Integrating computers in dual language programs includes having the computer used for both languages, with culturally and linguistically sensitive programs. The teacher can use the computer to interact and share experiences with parents more effectively . Through free and readily available group sites, teachers can share stories, pictures, permission slips and messages. With dual language programs new to many parents, knowing that children are learning in both languages is reassuring. Technology affords access to cultural references often unavailable to the classroom teacher. Art, stories and music are readily available. Rituals can be observed and imitated in play.

There is controversy surrounding the use of computers in early childhood classrooms. As with television commercials, young children are mesmerized by the fast moving images and electronic sounds of the computer games designed to engage them. There is no research on the long term effects of this high level of sensory stimulation has on children.

The computer within an early childhood classroom is simply one more choice where children may choose to work. A compelling concern with extended computer time by the group or an individual child is the experiences given up by computer time. Although different than television, in that computers are interactive and children can draw and make sound and motion happen, a computer program does not provide the same experience for example with balance and motion as a children have in the block corner. Although drawing on the computer may allow for representation it does not have the same kinesthetic properties.

It Is All About Quality

In both an English-only or dual-language program, the quality of the program and the individual classroom is what matters. To support quality, the revised *Developmentally appropriate practice in early childhood programs serving children from birth through age eight* (DAP) (Copple & Bredderkamp, 2009) provides empirically based principles and practices to promote quality in early childhood programs for all children and families. The DAP promotes culturally and linguistically sensitivity and inclusive practices appropriate (Copple et al., 2009) for dual language classrooms. By respecting and valuing culture and language, teachers promote and encourage the active involvement and support of all families (Copple et al., 2009). The DAP principles include:

1. Domains of children's development—physical, social, emotional, and cognitive—are closely related. Development in one domain influences and is influenced by development in other domains.
2. Development occurs in a relatively orderly sequence, with later abilities, skills, and knowledge building on those already acquired.
3. Development proceeds at varying rates from child to child as well as unevenly within different areas of each child's functioning.
4. Early experiences have both cumulative and delayed effects on individual children's development; optimal periods exist for certain types of development and learning.
5. Development proceeds in predictable directions toward greater complexity, organization, and internalization.

6. Children are active learners, drawing on direct physical and social experience as well as culturally transmitted knowledge to construct their own understandings of the world around them.
7. Development and learning result from interaction of biological maturation and the environment, which includes both the physical and social worlds that children live in.
8. Play is an important vehicle for children's social, emotional, and cognitive development, as well as a reflection of their development.
9. Development advances when children have opportunities to practice newly acquired skills as well as when they experience a challenge just beyond the level of their present mastery.
10. Children demonstrate different modes of knowing and learning and different ways of representing what they know.
11. Children develop and learn best in the context of a community where they are safe and valued, their physical needs are met, and they feel psychologically secure.

(Copple et al., 2009, pp. 5 –9).

Conclusion

Learning a second or more languages prepares children for the global society. If children do not learn a second language in early childhood, they learn the language differently. Children entering elementary school proficient in two languages are at an advantage as the second language supports the development of later literacy skills. As in all programs, early childhood dual-language classrooms need to maintain high quality.

Quality dual language program are language rich environments where children make choices from developmentally appropriate activities and materials, hear and express themselves in two languages as they play with adults and peers. Integrating technology into the early childhood classrooms assists teachers in dual language program by increasing communication with parents and by providing cultural and language materials that are not readily available in many communities.

References

Bruner, J. (1983). Play, thought, and language. *Peabody Journal of Education, 60*(3), 60-69.

Candelaria-Greene, J. (1996). A paradigm for bilingual special education in the USA: Lessons from Kenya. *Bilingual Research Journal*, 20(3 & 4): 545-564.

Christie, J., & Roskos, K. (2006). Standards, science, and the role of play in early literacy education. In D. Singer, R. Golinkoff, & K. Hirsh-Pasek (Eds.), *Play=Learning*: How Play Motivates and Enhances Children's Cognitive and Social-Emotional Growth. Oxford, UK: Oxford University Press.

Copple, C., & Bredekamp, S. (2009). *Developmentally appropriate practice in early childhood programs serving children from birth through age eight* (3rd ed.). Washington, DC: NAEYC.

Curtain, H., & C. Pesola. (1994). *Languages and children*, New York: Longman.

DeBey, M., & Bombard, D. (2007). Expanding children's boundaries: An approach to second language learning and cultural understanding. *Young Children Journal, 62*(2), 88-93.

Dickinson, D. K., & P. Tabors. (Eds.), (2001). *Beginning literacy with language: Young children learning at home and school.* Baltimore, MD: Paul H. Brookes.

Elkind, D. (2008). Can we play? *Greater good magazine: The science of a meaningful life IV. (4).* Retrieved June 2, 2009, from http://greatergood.berkeley.edu/greatergood/2008spring/Elkind344.html

Ervin-Tripp, S. M. (1974). "Is second language learning like the first?" In *Second language acquisition,* (E. M. Hatch, Ed.), 190-206. Rowley: Newbury.

Espinosa, L. (2008). Challenging common myths about young English language learners. *Foundation for Child Development Policy Brief; Advancing PreK-3*. New York: FCD.

Genesee, F., J. Paradis, & M.B. Crago. (2004). *Dual language development and disorders: A handbook on bilingualism and second language learning*. Baltimore, MD: Brookes.

Improving Head Start for School Readiness Act of 2007, Pub. L. No. 110-134 (2007).

National Association for the Education of Young Children. (1995) Technology and young children, 3 -8. NAEYC Position Paper. Retrieved June 2, 2009, from http://www.naeyc.org/about/positions/PSTECH98.asp

National Institute of Child Health and Human Development. (2000). *Report of the National Reading Panel. Teaching children to read: an evidence-based assessment of The scientific research literature on reading and its implications for reading instruction: Reports of the subgroups* (NIH Publication No. 00-4754). Washington, DC: U.S. Government Printing Office.

Nelson, K. (1996). *Language in cognitive development: The emergence of the mediated Mind.* New York: Cambridge University Press.

Petito, L.A., Katerelos, M., Levy, B.G., Gauna, K., Tetreault, K. & Ferraro, V. (2001). Bilingual signed and spoken language acquisition from birth: Implications for the mechanisms underlying early bilingual language acquisition. *Journal of Child Language*, 28, 453-496.

Piaget, J. (1962*). Play, dreams and imitation in childhood.* New York: Norton.

Sanchez, S.Y., & E.K. Thorp. (1998). Policies on linguistic continuity. A family's right, a practitioner's choice, or an opportunity to create shared meaning and a more equitable relationship. *ZERO TO THREE, 18*(6), 12-20.

Snow, C.E., M.S. Burns, & P. Griffin. Eds. (1998). *Preventing reading difficulties in young children.* Washington, DC: National Academy Press.

Stechuk, R.A., Burns, M.S., & Yandian, S.E.. (2006). *Bilingual infant/toddler environments: Supporting language & learning in our youngest children.* A guide for Migrant and Seasonal Head Start Programs. Washington, D.C.: Academy of Educational Development.

Tabors, P., & C. Snow. (1994). English as a second language in preschool programs. In *Educating second language children: The whole child, the whole curriculum, the whole community,* ed. F. Genesee. Cambridge, England: Cambridge University Press.

U.S. Dept. of Health & Human Services/ACF. (2001). *Head Start Program Performance Standards and Other Regulations*. Washington, DC: U. S. Department of Health and Human Services: Administration for Children, Youth and Families.

Vygotsky, L. (1966). Play and its role in mental development. *Soviet Psychology, 5,* 6-18.

Webb, S. (2007). Educating children in the new millennium: Child's play. *Education Week*, October 10.

Winsler, A., Diaz, R., Espinosa L., & Rodriguez, J. (1999). "When learning a second language does not mean losing the first: Bilingual language development in low-income, Spanish-speaking children attending bilingual preschool." *Child Development, Vol. 70*(2), 349-362.

Zebsko, N., & Antunez, B. (2000). If your child learns in two languages. *A parents' guide for improving educational opportunities for children acquiring English as a second language.* Washington, DC: National Clearinghouse for Bilingual Education.

Chapter 5:

Beyond Blackboard: An Update on Course Management Systems

Jeffrey S. Ruth

East Stroudsburg University

Currently, 82.3% of respondents have used a CMS, most of them using it several times per week or more often.

(Salaway, Caruso, & Nelson, 2008)

And have you tried Blackboard/WebCT recently? Sheesh. I mean, Sakai and Moodle are just sitting there...

(Dean Dad, January 9, 2009)

Despite a persistent vein of criticism from faculty and students, online education and blended learning modes continue to grow within the university sector. The latest survey from the Sloan Consortium (2007 data) reveals that 20% of all U.S. students are enrolled in at least one online course (Allen & Seaman, 2008). That percentage is more than twice the figure from five years before. Blended or hybrid learning is likewise on the rise (Picciano & Dziuban, 2007). Within the fastest-growing slice of higher education – community colleges – online enrollments recently rose 11% in a single year (Instructional Technology Council, 2008).

Naturally, a major component of this fundamental shift in learning is the vehicle itself, that is, the software used to serve up course content to students. Such software has long been dominated by Blackboard, a product whose very name has become synonymous with a radical reorganization of classrooms from chalk and slate to computers and networks. Today, however, Blackboard's leadership has weakened. As technological innovations continue to improve interactivity on the web – often for free via social networking services – open-source practices and standards have become increasingly accepted by end-users in higher education. Even as Blackboard has aggressively acquired two former competitors – WebCT and ANGEL – new and serious competition is emerging from open-source systems, especially Moodle and Sakai. These products now lead innovation in the field, and have produced a volatile market shift away from commercial course software.

Ultimately, the open-source movement and new social-learning technologies are calling into question the very role of courseware programs themselves, and pointing to a different scenario for future online learners. The purpose of this chapter is to survey the landscape of commercial and open-source courseware systems in use today – focusing especially on the dynamics of cost and features – and to describe the changes that are already leading faculty and students toward a new generation of e-learning environments.

Terminology

Settling on a single term to describe the dozens of software programs available to support online learning is difficult, if not impossible. Table 1 describes several of the terms most in use.

Table 1: Terminology

ABBREVIATION	MEANING	DESCRIPTION/USE
CMS	Content Management System	Software that allows a user to author content, delete and add it easily, and generally maintain a comprehensive site of text and multimedia; included here to distinguish it from another CMS (below) with a different meaning.
CMS	Course Management System	Software used to create and maintain an online learning environment for courses, including features like syllabus, multimedia resources, blogs, wikis, discussion forums, messaging, calendar and often grading.
VLE	Virtual Learning Environment	Term often used synonymously with Course Management System, especially beyond the United States.

LMS	Learning Management System	Software for delivering course experiences, including tracking course progress and organizing records, often in a corporate, enterprise environment, and often providing broader functions than those found in a Course Management System.

This terminology is often applied rather indiscriminately and casually among faculty and IT administrators, resulting in occasional confusion and a lack of clarity (Watson &Watson, 2007). In addition, most of the better known online software – i.e., Blackboard, ANGEL, Moodle – can be found in larger or smaller scale versions and installations, with differing features. As a result, these products are variously described as CMS or LMS (and often VLE beyond the United States). An example of the kind of relativity of terms applied to online learning systems can be seen in this description of the Interact 2.4.1 product: "Interact is an open source Learning Community Environment. It could also be called an LMS (Learning Management System), a CMS (Course Management System), or a VLE (Virtual Learning Environment)" (http://php.opensourcecms.com/). Similarly, at the Moodle site (http://moodle.org): "Moodle is a Course Management System (CMS), also known as a Learning Management System (LMS) or a Virtual Learning Environment (VLE)."

The kind of product being examined in this chapter is most widely known by the term Course Management System (CMS). Blackboard, Desire2Learn, Moodle, Sakai and other leading CMS systems do have capabilities that allow them to function more like an LMS (including large-scale tracking, enrollment analysis, etc.), but the focus here will be on usage that takes place at the course-level, typically led by a faculty member, and without institution-wide analytical functions such as student services, financial aid, etc.

Timeline

A few milestones in the development of CMS programs can help put into perspective the present situation. After the theoretical foundations for networked, automated learning were laid in the 1960s and 70s by R. Buckminster Fuller (*Education Automation*) and Ivan Illich (*Deschooling Society*), practical educational applications were attempted with the use of the emerging Internet. These included USENET, inaugurated in 1979, and the Open University Virtual Summer School (1994). A significant step forward came with the establishment of Nicenet (1995 and 1998), which provided a free Internet Classroom Assistant (ICA) to educators for distributing materials online and for facilitating communication among a class community.

The later 1990s also witnessed the development of the Bodington VLE – still with healthy usage in Britain today – and the release of WebCT (1997) and Blackboard (1998). Another 1998 release was the Ilias LMS project. In 2000 the ANGEL LMS was formed out of a prior university project (at IUPUI); in 2001 Desire2Learn entered the market; and in the following year Moodle was launched. The open-source Sakai Project was initiated by a consortium of universities during 2004.

Other products have also appeared during these years, including ATutor, Claroline, DrupalEd, Dokeos, Docebo, Joomla, Saba, SumTotal, eCollege and more. An unusual entry to this group is Epsilen (2007), which marries an online course environment to the entire online knowledge base of the New York Times. The emergence of these programs, collectively, has contributed to the exceptionally innovative CMS market we see today. Further, of the eighteen CMS products mentioned above, fully ten of them are free and open-source.

Shifting Market

Examining trends and market shares in the CMS sector requires a broader look at online education itself. Most researchers divide online education into four kinds: traditional (0% online); web-facilitated (1-29% online); blended/hybrid (30-79% online); and online (80%+ online). The latest Sloan Consortium study of online learning, *Staying the Course*, reports continued strong growth in this last category, as follows:

- Over 3.9 million students were taking at least one online course during the fall 2007 term; a 12 percent increase over the number reported the previous year.
- The 12.9 percent growth rate for online enrollments far exceeds the 1.2 percent growth of the overall higher education student population.
- Over twenty percent of all U.S. higher education students were taking at least one online course in the fall of 2007 (Allen & Seaman, 2008).

Another annual survey is sponsored by the nonprofit EDUCAUSE Center for Applied Research (ECAR). The *ECAR Study of Undergraduate Students and Information Technology, 2008* (cited on the first page of this chapter to highlight strong CMS usage among undergraduates) found that 93 percent of the institutions surveyed support at least one CMS (Salaway, Caruso, & Nelson, 2008).

From this data, an overall picture emerges to suggest that about one in five U.S. students currently takes a fully online course, and that about four in five students use a CMS, whether in web-enhanced, blended or fully online courses. The CMS has quite clearly become as standard a course tool as the chalkboard. At the University of Puget Sound, a committee noted the wide applications of a CMS (here termed LMS) across the disciplines:

> …the LMS has evolved from a simple repository for assignments, papers and the like, to providing grade books and discussion boards along with systems which allow the instructor to create and re-use sophisticated testing and assessment tools to evaluate and give feedback to student performance, allow students to post digital versions of assignments, to present ancillary digital material from textbooks, and much more. As our use of current technology becomes more sophisticated newer ways of using LMS are evolving, especially with the mass adoption of digital audio and video at the consumer level.
>
> For example Foreign Language instructors would like a way to electronically assess speech assignments and to provide an oral feedback. In fact, being able to vocalize feedback to a student may provide a more intimate and personal connection outside of classroom than say email or edited comments on a paper. The same is true of increasing faculty use of video: for foreign language homework, for Music auditions and practice review, for documenting the use of scientific lab instruments. Also, increasingly LMS play an active role in the classroom itself as the focal point

for presenting material to the class: teachers use it to bring up web links, show documents, have students take their quizzes in class, record audio exercises, and so forth. They have also adopted many of the tools of the "social software" revolution of web 2.0 with sophisticated forums, wikis, and blogs built into the course structure (LMIS Committee, 2008).

A wide variety of CMS products is available to accomplish these many tasks. A comprehensive list is published at the Edutools website, where approximately forty are featured and analyzed (http://www.edutools.info/index.jsp?pj=1). If one counts multiple products listed by single companies (e.g. Blackboard and ANGEL's several products), the total comes closer to sixty CMS's. Because of this situation, determining market share of CMS products themselves is at best an approximate science. A single institution might use more than one version of a product, and may officially support more than one company's products. Despite these complications (and as most higher education professionals already know), it can be safely asserted that Blackboard – together with the WebCT product that it acquired in 2005 – dominates the market. According to the annual *Campus Computing Survey* of 2008, 56.8 percent of responding campuses maintain Blackboard (or WebCT) as the only supported system (The Campus Computing Project, 2008). The commercial products Desire2Learn, ANGEL and Jenzabar are used at less than ten percent each. Two open source solutions Moodle and Sakai are used at 10 and 3.8 percent of campuses, respectively. Many other CMS products, including so-called homegrown ones, are also employed, but at a lower rate.

Although the 56.8 percent market share of Blackboard/WebCT is enormous, it has dropped steadily over the last four years. In 2002 that share was 76 percent, and as recently as 2007 it held at 67 percent. Its lost share during the last year has been picked up partly by a surge in the use of ANGEL at the community college level – more invested in online teaching than four-year colleges – and by Desire2Learn.

Most notably, however, the falling share of Blackboard/WebCT can be contrasted with the rise of the open source products Moodle and Sakai. Moodle is especially well represented in private, four-year colleges (e.g. Bryn Mawr, Carlton, DePauw, Lewis and Clark, Macalester, Reed, Smith) and in community colleges, while Sakai is slightly more represented in four-year public universities. A useful map of U.S. and worldwide Sakai usage can be found at http://sakaiproject.org.

This accelerating trend away from commercial CMS programs to open-source products is represented in Figure 1.

Figure 1: CMS Market Shift

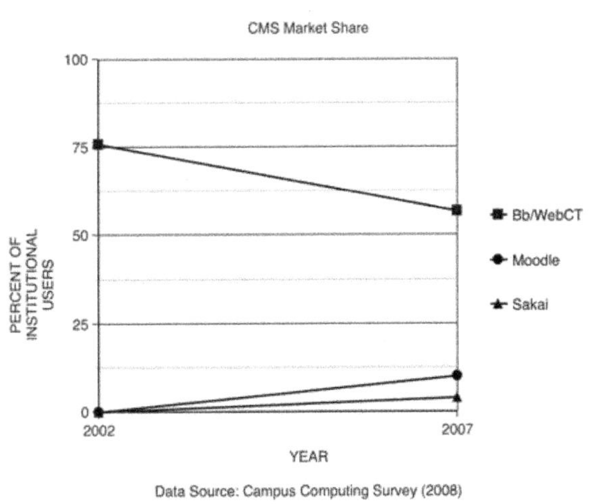

It is noteworthy that this trend is expected to endure and even intensify in coming years; the 2008 Campus Computing Survey reported that nearly one-quarter of institutions polled indicated a "high likelihood" that they would migrate to an open source program within five years (The Campus Computing Project, 2008). In the past, relatively few IT administrators had been willing to risk the go-it-alone route that open source programs appeared to present, given the high stakes game of keeping faculty and student course projects humming smoothly at all hours. Clearly, this is changing, and open-source adoptions appear destined to increase.

It is instructive also to note that in the United Kingdom, open-source CMS solutions are far more prevalent than in the United States. The UK's Open-Source Software Watch notes that

> Moodle was found in 6 out of 10 FE [Further Education] and a few HE institutions. The closed source Blackboard ran in a fifth of institutions, and another closed source solution – WebCT – had a similar percentage, but only in HE. In 2008, Moodle has grown significantly in HE, overtaking WebCT and it is catching up with Blackboard (it should be noted that Blackboard now owns WebCT, though). The fact that Moodle has spread from FE to HE is interesting, as it is usually assumed that HE institutions have the initiative with IT innovation and open source. But in the VLE domain it appears that FE institutions started to become involved, especially with Moodle, and now HE is following suit. Moodle is arguably a good example that given a clean slate, open source can be as successful, if not more so, than closed source. (Casero Cañas, 2008)

Why an Open-Source CMS?

Several factors account for the shift toward open-source course management systems.

Cost Dynamics

According to 2005 figures analyzed by Edublogs founder James Farmer, a common list price (initial fee) for a Blackboard system is over $300,000 (Farmer, 2006). Their product is sold with an annual licensing fee, as well. Like many software licensers, Blackboard charges approximately 20 percent of the sale price each year, resulting in the reported average of $75,000. In addition, Farmer reports that Blackboard typically pushes sales of peripheral components and services – its service business is very profitable – so that a university's initial cost and subsequent licensing fee can be far higher. More modest estimates of Blackboard fees can be had, but tens of thousands of dollars per year is not unusual for a rather basic license, before extra integration with IT services across the campus (e.g. Banner system) (Trotter, 2008). Much of this considerable cost to the university pays for activities that Blackboard assumes as a corporate entity, as explained by e-Literate blogger Michael Feldstein:

> Blackboard deploys a small army for each bid–not all of which result in a sale. Contrast this with Open Source. If you want to evaluate Moodle, you…well…you download Moodle. On top of the direct costs of sale, Blackboard has to sponsor conferences, support user groups, pay for fancy conference exhibit booths, and so on, because their customers expect them to do so (Feldstein, 2006).

It is certainly true that these enormous fees pay for service to the client, but free, open source products – especially Moodle – now boast vigorous forums worldwide, with the usual archived and searchable threads, as well as a knowledgebase, bug-tracker and ticket-response system for emailed inquiries. In addition, third-party service vendors (e.g. Moodlerooms, Remote-Learner, Classroom Revolution, and NewSchool Learning) are available to take Moodle administrators by the hand.

As noted at the Sakai site, a university choosing open source "…avoids the risks of vendor lock-in and escalating license costs." Given the financial crisis that erupted in 2008, universities will be increasingly choosing open source CMS programs over commercial ones in order to save money.

Interface

WebCT and Blackboard have recently made strides in this fundamental area, but suffer from years of relative neglect. Users of Bb/WebCT typically require more clicks to accomplish a given task than in Moodle (i.e., emailing users). For example, when CNET featured Blackboard in its April 2009 posting as one of "Five Tools for the World's Best Teachers," a plethora of protesting comments rained in. One stated, "I have used the WebCT line extensively and agree with other commenters that it lacks intuitiveness. Alternatives such as the open-source Moodle would certainly be worth considering over this" (Reisinger, 2009). Another described Blackboard's shortcomings: "I have to agree with you, as an educator who has used it extensively and been driven crazy by its poor, non-intuitive interface. It makes me wonder how much time Mr. Reisinger [the post's author] could possibly have spent with it. My wife, who is still a professor, continues to be plagued with its interface glitches and has to spend some of her class time teaching her students how to do what should be relatively simple things." By contrast, Moodle does not require the instructor to flip back and forth between his or her view and the student's view as the website is being built or modified. For such tasks, Moodle allows the instructor to simply make visible the editing buttons for all the features (e.g. forums, web pages, blogs, quizzes, etc.), which does not radically alter the basic view of the site.

Goshen College compared faculty satisfaction with the Moodle and Blackboard interfaces and found 47% preferred Moodle's, 27% Blackboard's (Saner, 2008).

As for Sakai, it has gained the reputation of a program whose complexities and depth sometimes impede usability. According to one blogger, "…the Sakai user experience has not kept up with other web 2.0 experiences, authoring content is awkward, and the program generally needs to be more flexible" (Rodes, 2008). Accounting for this, in part, is the reality that Sakai's code is written in Java, as opposed to the more accessible PHP code of Moodle.

Pedagogy and Inclusiveness.

Moodle is distinguished by a design that explicitly incorporates the principles of constructivism. This pedagogical approach emphasizes collaborative, reflective learning among all members of a course community. The core set of tools – forums, glossaries, wikis, assignments, quizzes, choices (polls), etc. – provides multiple avenues for communication among members. Capabilities for feedback and rating – qualitative and quantitative – are liberally laced into the program's features. This open, reflective approach has become a hallmark of Moodle's ever-growing feature set. In addition, the basic roles of administrator, teacher and student can be freely modified, mixing many different privileges in the way best suited to a particular course. Worth reproducing here are remarks made by Moodle's founder, Martin Dougiamas, on the spectrum of users that Moodle's structure needs to serve:

> At the fascist extreme there are those who want students to be highly controlled: reading resources that are revealed at set times and later sitting quizzes to prove they read those resources. I call this the rat-in-the-maze approach, or dump-and-pump.

> At the techno-hippy end of that spectrum there are those who want to devolve management completely, with every user running their own portfolio site, streaming blogs and files to each other using RSS and trackbacks. It's an interesting dream that really opens up thinking about education but I think the problems to be solved are many (such as security, accountability, the structure of institutions etc).
>
> The vast majority of people that this author encountered fall somewhere between these two extremes. Many of them are new to online learning, and are looking for the next step beyond what they were being paid to do offline, while being accepting of gentle guidance to improving their online techniques. These people are on a steep learning curve already without facing the aggregation of 100 different blogs (Dougiamas, n.d.).

This constructivist approach may be seen in a broader sense within the community of developers. Teamwork and rounds of review and self-examination form the core cycle of development, and mirror the pedagogical framework of the program itself.

Sakai is used and admired for many of the same reasons that Moodle is, and its basic feature set includes similar tools. Yet Sakai was founded within a far narrower frame of collaboration by several research universities, and its community of developers is guided by a board – the Sakai Foundation – that provides overall direction. Some Moodle users have noted Sakai's dependence on capital investment, with the risks that this carries to the process of an open source community (http://moodle.org/mod/forum/discuss.php?d=38661).

In a recent post ("Openness and Sakai History"), Sakai Foundation's Executive Director, Michael Korcuska, took on the perception of a closed process of development within the Sakai user and developer community. He states that in Sakai's first few years "…the Sakai community was either too closed or wasn't doing a good job in helping new folks get involved. We have improved on both of these fronts. We're also currently in the process of rebuilding the Sakai website (and maybe the wiki) with a strong emphasis on welcoming newcomers to the community" (Korcuska, 2008). In fact, fundamental change is moving Sakai's internal processes. In March of 2009 the Sakai Foundation issued a white paper detailing a new roadmap for incorporating its community, and for centralizing the development flow of new projects and features as it heads toward a major new release, Sakai 3. How will this affect the pedagogical experience of end users at universities around the world? Since the white paper explicitly notes that Sakai's current changes "will lead to the desired outcomes of an end-user driven set of development priorities" (Korcuska, 2009), it seems that more attention will be paid to the *desiderata* of faculty and students.

Blackboard has become the first choice among universities shopping for a CMS solution in large part because it claims to simplify the life of an IT administrator, performing many CMS tasks with one product. A related selling point is that allied Blackboard products can be purchased – for portal and content management, assessment tracking, advisement, accreditation and institutional planning – embracing many of the total IT needs of a university. Of course, this large-scale integration of functions is built not on a constructivist approach, rather a linear and hierarchical structure. While Moodle and Sakai emphasize the unity within a learning community, allowing for multiple avenues of student contributions, Blackboard tends to differentiate the roles and privileges of the participants (e.g. instructor and students). As noted by Lisa Lane in a

2007 article, "…the pedagogy of Blackboard/WebCT is based on managerial and administrative tasks centered on instructor efficiency: gradebook, test creation, threaded discussion, rosters, and instructor-student messages" (Lane, 2007). She notes that Blackboard/WebCT's default features (syllabus, assignment, announcement, etc.) are product-oriented and encourage the "sage on the stage" pedagogy. By contrast, other more constructivist CMS programs adopt the process-oriented, "guide on the side" posture. Novice online instructors tend to be satisfied with the basic options for posting static content onto their Blackboard/WebCT site (e.g. syllabus, assignments, text files, etc.), but they become more frustrated when attempting to weave more creative learning paths into these sites. George Siemens (University of Manitoba) has documented with great detail how adoption committees within universities often allow the pedagogical potential of a CMS to be overshadowed by its management features (Siemens, 2006).

Blackboard has more recently attempted to appeal to constructivist principles, as in this statement from its website (http://blackboard.com): "Effective communication is the cornerstone of effective education. Blackboard offers a suite of communication and collaboration tools that enhance student engagement and provide opportunities for social learning." In fact, after years of building market share and hence establishing the very nature of the CMS environment, the company has found itself suddenly in the role of follower, attempting to infuse its more rigid products with the inclusivity and constructivist mindset that have long overtaken K-16 pedagogy.

Modularity

The M of Moodle (Modular Object-Oriented Dynamic Learning Environment) is an explicit reminder of the collaborative, growth-oriented nature of the program. Currently over three hundred modules or plugins have been developed by contributors from dozens of countries. These modules (mostly Blocks or Activities) include Lightbox, Moogle, iTunes University Block, Sending SMS Block, Podcaster, EduGame, Moodle for Mobiles, and many others. Moodle has been able to incorporate technical advances in the web and the ever-changing social habits of web-users. Sakai has likewise aimed for similar agility, but (pending its announced changes) still lacks the robust, open community of developers and users that Moodle has. Sakai does boast an orientation toward features that assist institutions with assessment, accreditation and development of student portfolios (including Open Source Portfolio; see http://osportfolio.org).

Faced with a falling market share, Blackboard has seen the necessity of emulating the modular approach. Its site advertises the latest version, Learn (Release 9), by stating that "Tools must be easy to use and open to integration with other systems." While Course Cartridges and e-Packs have long been available – at a price – to supplement the product's functionality and to add content, recently the mostly-free Building Blocks have amounted to open-source add-ons. This clearly increases the utility of the product, and open-source advocates can only be pleased with this development. Still, some users are not convinced, such as this commenter to the Reisinger post: "What corporations like Blackboard do is try to guess where things are going – blogs and wikis for instance – and then figure out ways to lock down, monetize and render those tools useless" (Reisinger, 2009).

Next Steps

Here are some likely changes we will see in CMS programs and usage in the near and distant future.

More Open-Source Adoptions

The momentum of open-source CMS adoptions will increase. Recent adopters echo the motivations cited above, including pedagogical needs, feature preferences and cost savings. Regarding the University of Delaware's 2008 shift from WebCT to Sakai, Prof. Robert Simons of the University of Delaware noted, "I found that I could do all the things I'd done in the past, and that, in fact, some of the tasks were easier in Sakai than they had been in WebCT" (Gordon, 2008). Full support for WebCT by Blackboard will end in October of 2009, and this has provoked many current users into considering an open source adoption. The University of Puget Sound left Blackboard for Moodle in 2008: "Using an open source tool such as Moodle offers us benefits that go beyond the cost and flexibility issues. Being able to be a participant in how this resource is developed and used brings the faculty and even some students into the process. Having this be faculty/student-need driven gives us the greatest chance of success" (LMIS Committee, 2008).

Hosting providers for open-source CMS will continue to support these adoptions. Companies like Moodlerooms and rSmart provide off-campus hardware, free software (Moodle and Sakai, respectively) and 24/7 hotlines to institutions, all at a significant savings compared to commercial scenarios. For examples of cost analysis, see two sites: http://www.rsmart.com/savings and http://www.educause.edu/Resources/AnOutsourcedOpenSourceLMSandaP/162730. Details of such savings emerged in the 2007 public decision by Lousiana State University to switch from Blackboard to Moodle/Moodlerooms. The latter's annual cost was $85,000, compared to $330,000 for Blackboard, $300,000 for Desire2Learn and $260,000 for ANGEL (Farmer, 2007). As education blogger Dennis O'Connor states:

> What keeps budget conscious college administrators paying through the nose for Blackboard, D2L, or E-College? This author theorizes they are risk adverse. They stay safe and allow inertia to rule. By sticking with the big outfits that overcharge for online services they can 'play it safe'. Who ever lost their job by hiring Blackboard to provide an online learning environment? But what if it becomes clear that you could save a million dollars+ over a decade by going with Moodle? Hmm… […] Could someone in admin get canned for NOT being technically adept enough to make the Moodle move? (O'Connor, 2007).

Many other higher-ed adoption stories echo this one (e.g. UCLA's Common Collaborative Learning Environment). Large and small institutions who are tempted to reject the costly but 'safe' CMS providers and migrate to open-source systems will be nudged along by the broader shift among students and faculty toward social networking applications – e.g. Twitter, Skype, Flickr, Diigo – all available in robust, free versions.

Another factor promoting a shift toward open-source CMS adoptions is the parallel movement now underway in other areas of higher-ed IT, such as portals, document management, email service and social networking. An example of this is the array of free software chosen by Athabasca University (Canada; 37,000 students; 100% online). At Athabasca, Moodle is used alongside UPortal, Alfresco, Zimbra, ELGG, and several other major open-source products that have wide usage around the world.

More Integrations and Modularity

All CMS programs will inevitably expand their modularity in the coming years, following the model forged most notably by Moodle. Blackboard is developing open-source plug-ins that will actually allow bidirectional transfer of course content between itself and the two leading open-source CMS's. These two Building Blocks, called Moodle Connector and Blackboard-Sakai Connector, have already been greeted with skepticism by some open-source advocates, who believe that the plug-ins may not be truly two-way, notwithstanding claims to the contrary by Blackboard representatives. Others, the open-source optimists, have on their side the overwhelming evidence of an ongoing shift toward Moodle and Sakai, based on comparative advantages of these programs, and the likely scenario that these new bridges will make it easier for current Blackboard clients to finally experience open-source alternatives.

Another Blackboard module already deployed is the free application for the iPhone and iPod touch which allows students to access their accounts while away from their computers, receiving updates and alerts on grades, assignments, tests and other information.

Also seeking to ride the wave of modularity is Desire2Learn, which has developed a "2GO mobile learning application." ANGEL LMS, traditionally a leader in strong interface and modularity, now links to academic standards documents, provides a rubricmaker and lesson plan templates, and allows for easier embedding of multimedia and content mashups. Moodle already has some CMS integration with the Webctimport plug-in, and others.

Integration is also emanating from web-media companies. WebEx, a leader in live-conferencing, has designed a tool for using WebEx within the Blackboard environment.

The company will soon release a similar tool for use within Moodle (Cisco, 2009). Likewise, in 2008 Wimba created a module allowing its Wimba Collaboration Suite – with a variety of voice tools – to augment Moodle activities and blocks. A similar tool was already in use for Blackboard integration. Interestingly, Blackboard/Wimba offered this tool free to its K-12 clients. A striking new open-source alternative to web meeting tools is DimDim, now available freely within Moodle via the DimDim Web Meeting Activity.

Standardization of Protocols and Metadata

A key role in advancing CMS development around the world is being played by the IMS Global Learning Consortium (IMS/GLC), "a global, nonprofit, member association that provides leadership in shaping and growing the learning and educational technology industries through collaborative support of standards, innovation, best practice and recognition of superior learning impact" (IMS-Global Learning Consortium, n.d.). With worldwide members from major universities, software companies and educational nonprofit groups, IMS-GLC is coordinating the development and diffusion of universal standards in the area of online learning systems, including outcomes reporting, assessment and tracking (Learning Info Services, Question and Test Interoperability and Learning Tools Interoperability, respectively). These standards embrace transfer protocols, data format and much more. Collectively they comprise an emerging bundle of guidelines known as Common Cartridge, a more expansive and useful package of standards than the currently used SCORM (Sharable Content Object Reference Model), and will probably be significant for the future of the CMS landscape. Ultimately, Common Cartridge and other IMS initiatives are eliminating the difficulties of migrating from one CMS to another. Further, if most campuses are still Blackboard/WebCT clients, and many of these have declared a desire to migrate, the Moodle/Sakai alternative will loom as an increasingly desirable option, since compatibility problems and migration-cost will be minimal. Such uniform technical formats are already having their impact. When Cal State adopted Moodle and ANGEL in 2008, for example, migrating from one system to another was a relatively low-cost piece of the puzzle, eased as it was by interoperability standards in observance (Feldstein, 2009).

New Functions and Tools

Novel approaches to CMS programs are constantly being created, and more will emerge. The immersive, 3D online environment, Second Life, has by now acquired a critical mass of sorts among higher ed users. A free plug-in, Sloodle, allows Moodle users to seamlessly access their Second Life course activities from the Moodle course page.

Another major step forward for CMS products will be increased structure in a learner's training experiences. For example, in Moodle's version 2.0 (2009 release), a new feature called Completion Tracking allows for conditional activities, whereby a student cannot advance until a certain sequence of activities has been completed, with certain results.

It is likely that CMS programs will soon be providing students with a more portal-like experience, connecting their coursework with other areas of their computing life, including email, general file storage, etc. For example, the use of portfolios is growing exponentially in all aspects of academic life (e.g. in courses, for students' teaching licensure, for institutional accreditation, etc.). Blackboard, ANGEL and Sakai already take advantage of portfolio tie-ins (free or not), and smoother linkages will emerge. Moodle 2.0 will include a major step forward in this direction with its Portfolio module. Likewise, it will feature a Repository API that engages the Alfresco document storage capability.

It appears inevitable that the ubiquitous social networking sites (Flickr, Facebook, Twitter, et al.) will continue to make inroads to CMS feature lineups. The still amorphous semantic web, or Web 3.0, will surely be integrated into CMS tools (perhaps first in search options), with quicker appearance in open-source systems than in commercial ones.

Until recently, the most widely used open-source blogging software, WordPress, has been just that, and not a vehicle for wider courseware functions. Like Moodle, its extraordinary success has depended largely on an avid development community. In 2007, the ScholarPress group (http://scholarpress.net) created Courseware, a plug-in that allows WordPress to acquire several CMS functions. These currently include class schedules, assignments and bibliography. One example may be seen at the homepage of a course on Digital History at American University (http://clioweb.org/courses/digitalhistory/). Given the popularity of WordPress and the growing use of WordPress MU (multi-user) in higher education, we can expect to see more efforts to make WordPress emulate a CMS in the near future.

Beyond the CMS?

Are universities better off without course management software of any kind? Jim Groom, who has developed numerous open-source-inspired initiatives at the University of Mary Washington (UMW), foresees a future built not around the coursepage but around the student. Currently, a student typically produces a variety of course-based resources – a links list, course blog posts, wiki contributions, multimedia presentations, etc. – that often collapse when the course page expires at semester's end. Instead, a student could be provided with a single, multipurpose homepage that can be linked to all current and past coursepages. Groom employs the WordPress MU platform to provide each of his university's 4000+ students with a page, and students customize and maintain this space according to their needs and motivation. How do they remain connected to their online course experiences? Through a syndication hub made possible through tags and RSS feeds wherein the work done by students in their own blog space is immediately re-published within a course aggregator page (Groom, 2009a). This model allows a student member to post and otherwise contribute to the "coursepage" *from his or her personal page*, and not at the coursepage itself. In this way, at semester's end, the student's work remains in the personal page, having been fed to the coursepage throughout the semester.

Groom and others refer to this student-centered arrangement as distributed individual publishing. Naturally, all students might be expected to participate in certain activities via the coursepage (assignment submissions, assessments, synchronous chats, etc.), but the CMS as a whole would become far more de-centered and distributed. The conceptual shift is striking, since the pole of a student's online life shifts from several discrete coursepages (Blackboard, Moodle, WordPress, etc.) to his or her own page, with its multiple links to courses, portfolios, social networks, friend and family life, etc. Coursepages would still exist, but their role would be less central. A student would use his or her page throughout the college years and beyond, changing its shape and accumulating resources as study plans and goals moved forward. This mode of education gives more autonomy and responsibility to the student, but does not necessarily create isolation, since the learning still takes place in a socially networked fashion. John Seely Brown and Richard Adler provide additional detail on this (2008).

It's worth noting that this vision would require two key ingredients not in the present courseware landscape: (1) CMS programs would need to make the feed-aggregator architecture (described above, with WordPress MU) easy to use, and (2) faculty need to consistently adopt this student-centered model when designing course experiences. Groom, who proposes that all of this will become increasingly common on campuses as the current CMS model evolves and fades, provides this vision:

> De-centered, distributed publishing is a flow of information we are not used to. It's anarchic, somewhat confusing, and difficult to follow unless we are in the "natural" flow of things. Yet, that is the key here, once in the stream (and the idea of a stream here is a much larger aesthetic and design shift that the internet has been undergoing and reflecting more broadly for a number of years now) the trace of the arguments, discussion, and ideas become that much more naturalized to the movement of information within a community while at the same time keeping the power of design (think themes and widgets here) as well as the overall control of the space in the hands of the individual. (Groom, 2009b)

This new focus on student-as-hub is even more relevant when one considers the student as a learner who partakes in opportunities at multiple universities or portals (e.g. MERLOT). Freely available course content – Open Education Resources (OERs) – is expanding, led by the highly visible open courses at MIT, Yale, the UK's Open

University and several others around the world. As private and public institutions and individuals make OERs increasingly common, the single-courseware model for a learner becomes ever more obsolete. Instead, the Personal Learning Environment (PLE) is emerging, through which learners can take advantage of natural, symbiotic connections while browsing content from many courses simultaneously. The UMW approach with WordPress MU represents a step toward a PLE for students; an EDUCAUSE document comments that "The [UMW] system is flexible enough to allow students to present internally developed content side by side with work they cultivate and maintain elsewhere on the web. Similar approaches have been used at Baylor University, Penn State, and the University of British Columbia" (EDUCAUSE, 2009).

Open-source coders have begun to facilitate PLEs, creating tools that allow a CMS to share content with guest learners. A recent Moodle plug-in, Openshare, is described by its author, Jared Stein, like this: "The OpenShare block for Moodle provides controls for setting one or more resources or activities in a course as closed (private, for enrolled class members only) or open (available to the public). This allows you to use Moodle as a publishing platform for opencourseware or open educational resources" (Stein, 2008). Enrolled students – and the professor – would retain autonomy and privacy according to the needs of the course, while allowing for the serendipitous learning patterns of visitors.

Finally, university IT departments are also considering Google Wave (http://wave.google.com) for its potential role to facilitate online education. Announced with much fanfare on May 27, 2009, the initial, limited release of Wave allows a browser to reorganize emailing and instant messaging into a more seamless unit of communication, with content of all kinds added as needed by the users. Google designed this as a communication platform that enables collaboration and multimedia publishing of the sort that is now found in many CMS-based courses, so it is reasonable to imagine that this will be piloted immediately at some colleges as an alternative to the official, campus-supported CMS. In addition, Google will soon release the code to Wave as open-source software, a move that will propel the product into the hands of the developer community and promise it a fast-moving future. Ed-tech bloggers are already posing the question, "Is this the future of the CMS?"

Caveat

To imagine a future scenario of online education with Waves or PLEs, but not CMSs, may seem alarming to many instructors and students who are just now getting used to their own web-enhanced, CMS-based courses. Most faculty members, like the general public, are little prepared to begin thinking about Web 3.0 at a time when Web 2.0 is just starting to become familiar terrain. In fact, some are still asking, "What exactly was Web 1.0?" At a time when so many web-users are firmly committed to looking ahead, it seems important to glance around and recognize that not everyone is passionately embracing the next great wave of change.

While it seems certain that learning will become ever more computerized and networked (via CMSs, PLEs, Waves or some other future scenario), the ECAR 2008 survey noted multiple veins of student criticism toward their CMSs:

> Four themes emerged—that the lack of face-to-face interaction detracts from learning, that online courses facilitate cheating, that technical problems still exist, and that online courses require students to "teach themselves," making the courses more demanding.

One student was quoted with a paradigmatic complaint: "Nothing can adequately replace face-to-face lectures and recitations. Nothing. I don't care how expensive the computers are, how high-definition the video is, or how fancy the presentation software is." Other evidence in this regard is not lacking. A Facebook group member posted "All five of my classes use Blackboard and I hate it!! I miss paper hand outs and teachers actually TALKING to their students!" (Thomsen, 2009).

Two researchers from the University of Central Florida catalogued hundreds of student responses to blended learning courses, using semantic analysis to find strong ambivalence about the experience. Terms used by students included *double-edged sword, on the other hand, balancing act, both sides of the fence, gray area,* etc. (Moskal & Dziuban, 2009).

It seems probable that a student who thoroughly dislikes a CMS experience will remain displeased regardless of the software used. Anecdotally, a Google search of "I hate Blackboard" yielded 130,000 hits, and "I hate Moodle" no fewer than 30,300. Accounting for the ratio of users for each program, this informal indicator shows these programs to be more or less equally unloved. For these students – a minority, it must be said – there is humorous relief in the YouTube video "Blackboard or Moodle at UC," a parody of the dispute itself. It concludes with a rendition of the classic tune, "I can do anything you can do better" (http://www.youtube.com/watch?v=jYcaalHjsI8).

Student complaints of the sort described above – both casual and more substantial – are real. Because such attitudes correlate with lower academic success, they should serve as counterpoint or caution as CMS usage grows in higher education, and must be taken into consideration by professors when designing learning options for their courses.

Conclusion

In the end, the ongoing evolution of course management systems will certainly provide more satisfying formats for online learning, producing fewer notes of discontent among learners, instructors and IT administrators. As described above, current trends and technical developments suggest clearly that these future e-learning environments will be based on open-source software and openly-shared course content. Both of these online ingredients will be tethered to the unique demands and motivations of individual instructors and students, whose experiences will be increasingly enhanced by their own social learning networks, both in and out of the formal, online class setting.

References

Allen, I. E., & Seaman, J. (2008). *Staying the course: Online education in the United States 2008.* Sloan Consortium. Retrieved May 1, 2009, from http://www.sloanconsortium.org/publications/survey/pdf/staying_the_course.pdf

Athabasca University. (2008, June 19). Open Learning Environment. Athabasca and Open Source Infrastructure [presentation]. 2-3-98 Moodle Conference, Delhi, NY. Retrieved May 1, 2009, from http://moodlerooms.com/resource-center/case-studies/athabasca-university/

Brown, J.S., & Adler, R. P. (2008). Minds on fire: Open education, the long tail, and learning 2.0. *EDUCAUSE Review*, *43*(1), 16–32. Retrieved May 1, 2009, from http://www.educause.edu/EDUCAUSE+Review/EDUCAUSEReviewMagazineVolume43/MindsonFireOpenEducationtheLon/162420

Casero Cañas, R. (2008). *OSS watch national software survey 2008*. Open-source software watch. Retrieved May 1, 2009, from http://www.oss-watch.ac.uk/

Chao, I. T. (2008). Moving to Moodle: Reflections Two Years Later. *EDUCAUSE Quarterly*, *31*(3), July-September. Retrieved May 1, 2009, from http://www.educause.edu/EDUCAUSE+Quarterly/EDUCAUSEQuarterlyMagazineVolum/MovingtoMoodleReflectionsTwoYe/163101

Cisco. (2009, April 27). Enhanced communications. *Cisco.* Podcast retrieved from http://newsroom.cisco.com/dlls/podcasts/ciscocast_lee_042709.html

Dean Dad. (2008, January 9). Selective outsourcing. Message posted to http://www.insidehighered.com/blogs/confessions_of_a_community_college_dean/selective_outsourcing#Comments

Dougiamas, M. (n.d.) *Pedagogy.* Retrieved May 1, 2009, from http://docs.moodle.org/en/Pedagogy

Dralle, A. (2007). Analysis of blackboard and open source course management systems in online instruction. In T. Bastiaens & S. Carliner (Eds.), *Proceedings of World Conference on E-Learning in Corporate, Government, Healthcare, and Higher Education 2007* (pp. 7064-7066). Chesapeake, VA: AACE.

EDUCAUSE. (2009, May 5). *7 Things you should know about personal learning environments.* Retrieved May 1, 2009, from http://www.educause.edu/Resources/Browse/PLE/31814

Farmer, J. (2006). *On the cost of selling an enterprise learning system.* Retrieved May 1, 2009, from http://www.cedma-europe.org/newsletter%20articles/misc/On%20the%20cost%20of%20selling%20an%20Enterprise%20Learning%20System%20%20(Jan%2006).pdf

Farmer, J. (2007, November 5). Louisiana state university moves to Moodle. Message posted to http://mfeldstein.com/

Feldstein, M. (2006, January 13). Blackboard by the numbers. Message posted to http://mfeldstein.com/

Feldstein, M. (2009, November 19). CSU picks Moodle and ANGEL. Message posted to http://mfeldstein.com/

Gordon, R. (2008, March 20). Campus prepares for new learning management system. *University of Delaware Daily.* Retrieved May 1, 2009, from http://www.udel.edu/PR/UDaily/2008/mar/sakai032008.html.

Groom, J. (2009, May 11). Approximating the syndication bus with wordpress.com. Message posted to http://bavatuesdays.com

Groom, J. (2009, June 7). The design of openness. Message posted to http://bavatuesdays.com/

IMS-Global Learning Consortium. (n.d.) *Interoperability...Innovation...Adoption...Learning.* Retrieved May 1, 2009, from http://www.imsglobal.org/aboutims.html

Instructional Technology Council. (2008). *Distance education survey results. Tracking the impact of elearning at community colleges*. Retrieved April 30, 2009, from http://www.itcnetwork.org/file.php?file=%2F1%2FITCAnnualSurveyMarch2009Final.pdf

Korcuska, M. (2008, May 20). Openness and Sakai history. Message posted to http://sakaiblog.korcuska.net/

Korcuska, M. (2009). *2009 Sakai development process. Sakai board of directors recommendations.* Retrieved May 1, 2009, from http://mkorcuska.files.wordpress.com/2009/03/sakai-development-process-2009-v05b.pdf

Lakhan, S. E., & Jhunjhunwala, K. (2008). Open source software in education. *EDUCAUSE Quarterly, 31*(2), April-June. Retrieved May 1, 2009, from http://www.educause.edu/EDUCAUSE+Quarterly/EDUCAUSEQuarterlyMagazineVolum/OpenSourceSoftwareinEducation/162873

Lane, L. (2007). *Course management systems and pedagogy*. Retrieved May 1, 2009, from http://lisahistory.net/pages/CMSandPedagogy.htm

LMIS Committee. (2008). *Learning management system recommendation.* Retrieved May 1, 2009, from the University of Pudget Sound Web site: http://projects.ups.edu/rwthornton/LMS_Moodle_finaldraft.pdf

Moskal, P., & Dziuban, C. D. (2009). Reactive behavior, ambivalence and the generations: Emerging patterns in student evaluations of blended learning. Archived online seminar accessed on April 28, 2009 from http://www.blendedteaching.org/Blended_Online_Seminars

O'Connor, D. (2007, November 17). Moodlerooms saves ULM $100,000 a year. Message posted to http://doconnor.edublogs.org/

Open Source CMS. *Open source learning management.* Retrieved May 1, 2009, from http://php.opensourcecms.com/scripts/show.php?catid=6&cat=Learning%20Management

Picciano, A. G., & Dziuban, C. D. (Eds.). (2007). *Blended learning: Research perspectives.* Newburyport, MA: Sloan Consortium.

Reisinger, D. (2009, April 21). Five tools for the world's best teacher. *CNET News.* Retrieved May 1, 2009, from http://news.cnet.com/8301-17939_109-10223119-2.html

Rodes, R. (2008, December 18). Sakai 3. Message posted to http://rodeworks.com/

Salaway, G., & Caruso, J. B., with Nelson, M. R. (2008). *The ECAR study of undergraduate students and information technology, 2008* (Research Study, Vol. 8, Boulder, CO). Retrieved on May 1, 2009 from the EDUCAUSE Center for Applied Research Web site: http://www.educause.edu/ECAR/TheECARStudyofUndergraduateStu/163283

Saner, F. E. (2008, June 19). Learning management. A case study of critical issues [presentation]. 2-3-98 Moodle Conference, Delhi, NY. Retrieved May 1, 2009, from http://www.moodlerooms.com/resource-center/case-studies/goshen-college/

Siemens, G. (2006). *Learning or management system? A review of learning management system reviews.* Learning Technologies Centre, University of Manitoba. Retrieved May 25, 2009, from http://ltc.umanitoba.ca/wordpress/wp-content/uploads/2006/10/learning-or-management-system-with-reference-list.doc

Stein, J. (2008). *OpenShare for Moodle.* Retrieved May 1, 2009, from http://flexknowlogy.learningfield.org/addons/openshare/

The Campus Computing Project. (2008). *The 2008 campus computing survey.* Retrieved May 1, 2009, from http://www.campuscomputing.net/survey-summary/2008-campus-computing-survey

Thomsen, B. (2009, February 10). Message posted to http://www.facebook.com/group.php?gid=2246308162

Trotter, A. (2008, June 9). Blackboard vs. Moodle. *Education Week.* Retrieved May 2, 2009, from http://www.edweek.org/dd/articles/2008/06/09/01moodle.h02.html

Watson, W. R., Watson, S. L. An argument for clarity: What are learning management systems, what are they not, and what should they become? *TechTrends, 51*(2), 28-34.

Chapter 6:

Best Practices: Practicum for Classrooms as Learning Communities

Fabio Girelli-Carasi

Introduction

"Best practices" are a moving target, not a set of skills one can master by "practicing," as the name would misleadingly suggest. They are general directives that can become quite detailed and specific in the operational phase, but are not immutable. They do not strictly depend on the availability of technology either. The high-tech options for archiving, for instance, can be a centrally located file server, or a wireless home network connected to NAS (Network Attached Storage.) But best practices for archiving documents can be pursued in a low-tech fashion simply with a filing cabinet and manila folders.

When we limit our analysis to technology, several variables depend on the kind of hardware and software we rely on. As technology changes, so the *outcomes* of best practices also change. This may give the impression that best practices are just a fancy name for the "flavor of the month" approach to innovation in the field of education, whereby every once in a while a new "method" is introduced with the promise of being the magic bullet that will solve all problems in one swoop.

Technology itself, despite considerable hyping, will not, in and by itself, solve any problem, nor will best practices. What they can do, combined together in a meaningful, coherent approach, is to provide teachers both with a solid philosophical and structural grid and a nimble, effective set of strategies worked out in finite details, and precise suggestions on how to prepare the future citizens and intellectual leadership of tomorrow's world.

It will appear strange that while technology has the effect of liberating creativity and the assertion of individuality, at the same time it imposes extreme constraints on its very use. It is the paradox of technological progress. We gain something, but we lose something else. Let's take writing, for instance. To this very day we can read 5,000 years old inventories of the food warehouse of Babylon's royal court, with details on the harvest size of wheat, barley, millet, grapes and olives. They were written on clay tablets that ensured they would last for millennia. When writing switched to papyrus, it certainly gained in portability, but lost in durability.

Skipping a few thousand years, when humanity went from writing with a pen to a typewriter, we gained in speed and clarity but we lost control over space: we could no longer shrink the size of our handwriting to make words fit in a small space, nor could we add a quick note on the margin (they had to invent Post-It notes for that.) We also limited ourselves to a certain surface (paper) with predetermined size and thickness.

Moving from the typewriter to the computer, we are facing the fact that even before we can begin to write a word we have to follow a set of very rigid procedures: turn on the computer (which has to be plugged into a source of electric power), launch an application, name a file, learn all sorts of commands, save the file and finally print it with a different machine, to operate which we have to learn a whole new set of commands and procedures. The procedure is still so rigid that to this very day most people do not use computer and printer to write an address on an envelope. They either use an old fashioned typewriter, or, accepted practice even in business, they write it by hand.

How does this apply to teaching? One of the hardest things for a teacher is to get students to comply with instructions on protocols and formats. There is an anarchist streak in the students, philosophical anarchy, to be more precise. They seem to resent directions and requirements as if they were attacks on their creativity and freedom of expression.

Yet, the very first rule for a teacher who wants to use technology is exactly this: be very clear about protocols, be inflexible, train your students to abide by the rules. Do not compromise on the fundamental points of form and format. If you are tentative, you will be confronted with individuals who impose their preferences and idiosyncrasies on your very limited resources of time, energy and patience.

Best Practices: Communication and Communities of Learners

This chapter focuses on best practices as they pertain to a specific aspect of educational technology, namely *communication* and the development of *communities of learners.* This is distinct from the *instrumental* use, where technology serves as study-aid or for the creation of instructional materials.

This distinction turns out to be absolutely crucial. Technology is split in two distinct moments. The first moment is the evolution of several older technological apparatuses into the computer, namely a single, more complex and more productive one. Let's take, for instance, slides, in the recent past used in art history courses; audio cassettes for foreign language in language labs; VHS videos; pocket calculators and even transparencies and paper hand-outs. All these tools can be replaced by computers and the Internet, and all their functions can be performed by a single machine.

The second moment comes when we take all these discreet functions and we view them holistically, within the entire field of technology as a whole. This motley assembly of parts becomes a full fledged, self-standing *medium* that produces new forms of communication, and in so doing, applies to new forms of community.

When we apply this view to education, the first consequence we observe is that, maybe for the first time in the modern era, there is the realistic possibility to create *life-long learners.* This forces us to review and re-think education from the bottom up, and brings us closer to the ideals of thinkers and philosophers of education such as Maria Montessori and John Dewey, who posited lifelong learning and communities of learners as the highest possible achievements of education. These ideals have provided the theoretical and conceptual basis for new proposals of reform in education, so that society will be able to prepare the new generations for the *Era of Knowledge*, a term coined by Peter Drucker (1994) and adopted by other leading scholars of complex systems.

This chapter introduces some of the techniques that, taking advantage of the opportunities afforded by technology, allow us to structure an entire program or curriculum based on the involvement of learners as a community. In addition, we listed a number of prescriptions and "do's and don'ts," based on the experiences of practitioners and experts in the field. Some of these prescriptions will appear to be almost risible in how basic they are. However, they are the foundations of the entire structure and the essential guidelines without which teachers and students will find themselves wandering in a technological wilderness without any sense of direction.

The Introduction mentions the need to establish firm protocols and to enforce them rigorously. The clearer the expectations and the instructions, the sooner the students will comply. Clear also means streamlined and consistent.

Below is a partial list of areas where best practices have been developed.

1) Communication Protocols: Email.
2) Peer Evaluation, Standards and Critical Thinking.
3) Collaborative Activities: Research, Joint-Learning, Group Writing. Email, Blogs, Wikis, E-Portofolios, Social Networks.
4) Crowd Sourcing:
    a. Student-Generated Materials.
    b. Student-Generated Curricula.

With all manuals, on the part of the reader there is a tendency to apply the equivalent of the "law of gravity," that is to assume that the first chapters contain the easy information and that the further one advances, the harder the content becomes. When it comes to technology, my favorite visual metaphor is a series of concentric circles divided into sectors. Each component, represented by a different slice, is needed to complete the circle. At the core are the essential utilizations of each instrument. As we move outwardly, each subsequent circle indicates a broader application of the tools and a higher degree of integration into the curriculum (and complexity of the applications.)

Figure 1: Technology Tools

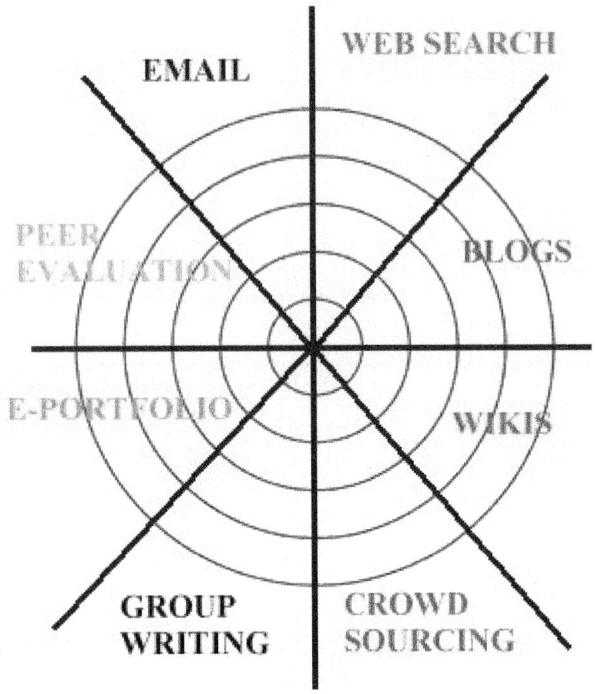

1. Communication Protocols: Email

The first tool in the technology arsenal that teachers will probably employ to communicate with students, and to help students communicate with each other, is in all likelihood *email*. This is an extremely powerful tool and, based on personal experience, at the very beginning it is all is needed for a teacher to start transforming a class into a community of learners. One does not need to learn the arcane complexities of html or even more complicated programming languages. All is needed is a *series* of email accounts.

To manage effectively the stream of messages, the first rule is to create a separate email account per class. This will allow keeping personal business separate from work. Moreover, it will prevent confusion about which student is in which course. If your institution does not provide multiple accounts, you should use a free online email service, such as *Yahoo Mail (*https://login.yahoo.com) or *Google Mail* (http://www.gmail.com).

The course address must be transparent and intuitive to the students, something like:

philosohy1-spring2010@.....

These email accounts can be abandoned or recycled. In the second case, of course, it is advisable to archive all the information from the previous semester and start with a blank slate.

The communication protocols for students must state as an absolute requirement that everyone must have a *professional email address*. This means an address that contains first and last name. If the address is already taken, students can add a couple of digits at the end to make it unique. The First and Last name must be separated by a *hyphen*, an *underscore* or a *dot* to make them easily legible.

Examples:

John.Smith@.....
John-Smith@.....
John_Smith@.....
John.Smith88@.......
Etc.

Encourage your students to use this address for all professional contacts, with other teaches as well as when registering for official business, such as the Motor Vehicle Bureau, SAT's and so forth. This is also the kind of address that should appear on their C.V.'s.

In addition to a professional address, it is important that the *screen name* is clearly spelled out. The "screen name" is the first information you see when you receive a message, in the "From" column of your email program.

Figure 2: Screen Names as They Appear in an Email Client Application

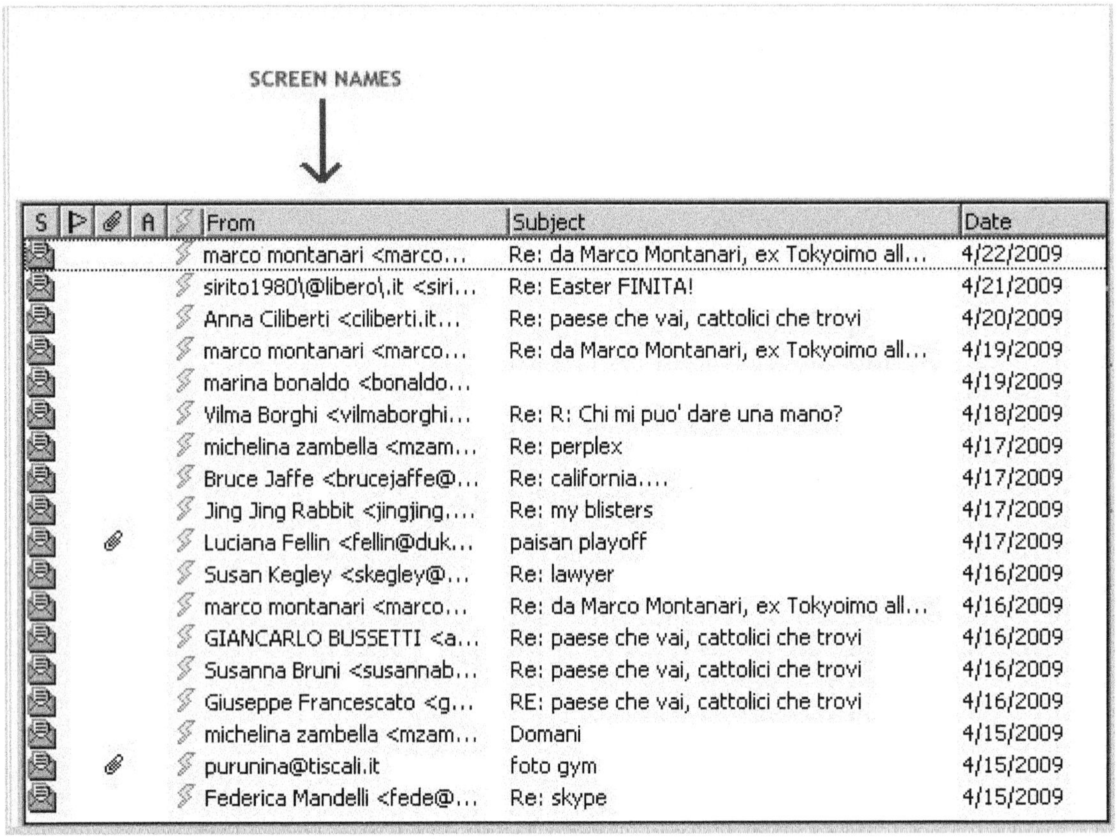

Here is how an opened message should look like:

Figure 3: Screenshot of Email Message (Headings)

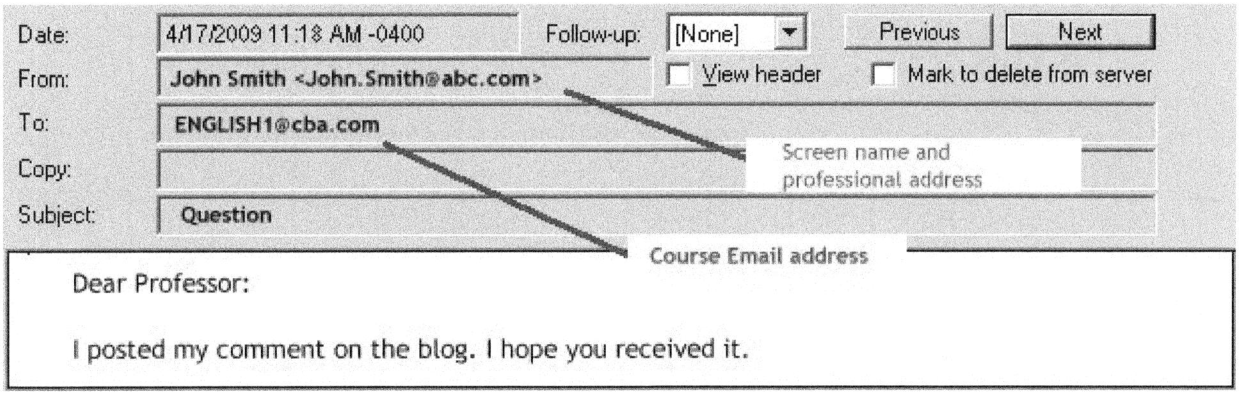

Third, essential requirement: the email account must be *accessible on the Web*. This is a very important preventive measure for self-defense. If (*"if?")* students claim they sent a particular message and blame "the Internet" for not delivering it to your mailbox ("the server ate my homework,") you can ask them to show you evidence by logging on to their account and open the "Sent mail" folder. (When you let them know this is the reason why you require a web-accessible account, you will not hear the 'the-Internet-ate-my-message" excuse ever again in your life.)

Below is a chart with recommended email protocols.

The Golden Rules of Email Correspondence

· Check your email daily.

. Exchange your email address with several classmates.

· Email is a <u>formal</u> means of communication (at least when you correspond with your professors or employers, etc.). Always be respectful. Use capital letters, punctuation, greetings and salutations as in a professional message. Avoid abbreviations, slang and cuteness. <u>Improperly formatted messages will</u>

be rejected or discarded without notice.

· The use of magic words (please, thank you) and politeness in general is strongly encouraged.

· Sign every message at the bottom. Why? Because it is good manners and if it is a long message, the reader doesn't have to scroll up to check who the sender is.

· For homework by email, the instructor will give you the exact words to be typed in the SUBJECT line. This allows the instructor to route messages automatically to the appropriate archival folder.

. If your homework is in an attached file, you will use the exact filename provided by the instructor. This allows the recipient to route your files automatically to the appropriate archival folder.

· Please do not write to let your instructor know that you were not in class (s/he already knows) or that you will miss class; to ask for a summary of what we did or will do in class; or what the assignment is/was. Email a classmate instead.

Practicum 1

i)

First task to assign to the students: "Send a message to the course's address." In order to make it a little more meaningful, the students will write a short autobiography in the email body. (It is worth repeating to the students that, as the old cliché goes: “You don’t get a second chance to make a first impression.” This usually motivates them to make an extra effort to show what they can do.) The teacher responds acknowledging receipt with suggestions for changes if the sender’s address does not comply with the specifications. This is also the most efficient way to capture all the students’ addresses.

ii)

To introduce the students to the notion that technology is part of everyday life, try using it *to take attendance*. It also serves the purpose of cutting down on lateness. At the very beginning of class, open your email account and compose a new message. In the address line enter the names of the students and a message as in the image below.

Figure 4: Taking Attendance by Email

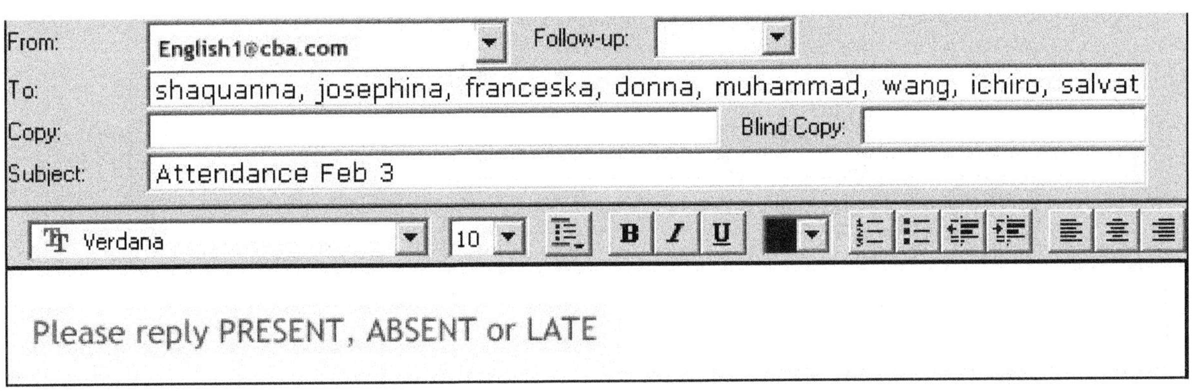

iii)

A good practice is to *assign homework* only and exclusively via email always using the same protocols and conventions:

Here is an example:

Figure 5: Assignment by Email

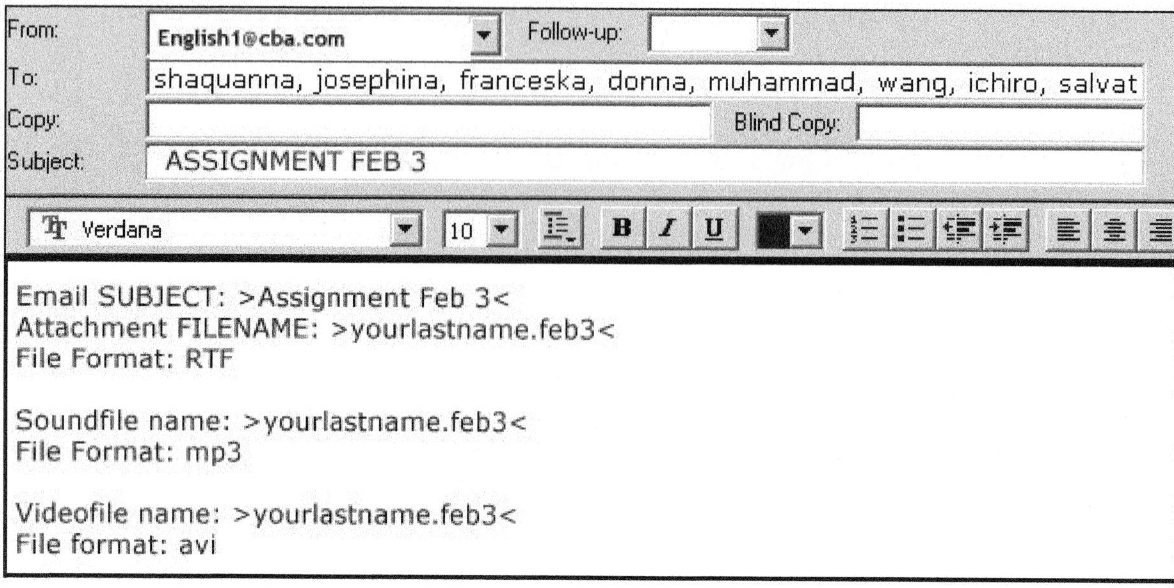

1) In the ***Subject*** simply state the content of the message and the due date of the assignment.

2) In the body, ***Email Subject*** is the information the students have to type in the Subject Line of their email message.

3) *Filename* is the name of the document containing the assignment, to be sent as an *attachment*. The filename requirements are extremely important in order to keep track of the students' work. It is composed simply of the *student's name* and the *assignment's due date.*

4) *File format*: students may use different word processing applications (Word, WordPerfect, OpenOffice, etc.) running on different platforms, such as Macintosh, Windows and Linux. To avoid compatibility problems, particularly vexing with the introduction of Microsoft Vista and subsequent generations, it is expedient to require a simple, universal format. RTF (Rich Text Format) is just the perfect solution

for word processing. Also require a specific format for soundfiles (suggested: mp3) and for video files.

It is imperative to insist that the students comply with these rules. One reason is purely practical: it makes the teacher's life easier. Let's not forget that technology adds complexity to our mundane tasks. We have an investment in ensuring that we don't let a data overload overwhelm us. If students submit an assignment with the wrong filename, or in the wrong format, at first they will receive a warning. If they persevere, the instructor should simply ignore the submissions. The second reason is to impress upon the students that technology requires attention to detail and acceptance of basic procedures and respect of protocols, as does driving a car or even walking down the street.

2. Peer Evaluation, Standards, and Critical Thinking

We lament the fact that very little communication actually takes place in a classroom, particularly among students. One of the most serious consequences is a generalized unawareness of the *evaluation standards*: plainly put, students have only a vague notion of what they have to do in order to get an A, or simply to pass the course. When the evaluation takes the form of multiple-choice tests according to rigid parameters, it's rather easy to figure out the percentage-based grading scale. At the other extreme, when a course is graded "on a scale" all bets are off. In between lies the majority of courses for which the evaluation by the teacher is a mix of fixed parameters and relative measurements. Depending on the overall quality of the class, teachers are subconsciously prone to evaluate individual performances more or less severely. These cases of "hybrid evaluation" are so frequent as to be almost the norm rather than the exception.

Aside from the personal, impressionistic factors, even objective parameters are often simply impossible to describe and define. As a consequence, the students, who rarely if at all share their work with each other, let alone the grades, live in a sort of individual cocoon, uninformed on the criteria used to judge their work and deprived of the necessary information needed for improvement. This is particularly acute in the humanities and the social sciences where the evaluation is often applied to written texts (writing assignments and papers.)

Students, in other terms, are unaware of the *performance standards* in their classes, and furthermore are unaware of the *performance expectations* by the instructors. By making the students aware of these standards we will also begin to unfold a strategy to help them develop the *critical thinking skills* that have become a sort of Holy Grail of modern education.

It is a personal opinion that these aspects of the educational process are tightly connected and interdependent. We cannot help the students develop *critical thinking skills* unless we provide a *body of knowledge* on which these skills can be practiced. The body of knowledge, in turn, cannot be any amass of notions and facts that were acquired previously or being acquired in the course. The body of knowledge resides in the student body, *in the students themselves*, and in their ability to observe their own *learning processes*. Within the domain of the learning process, we can teach our students the tools of *self-reflection*. From their use, *self-awareness* will result. The combination of these moments will yield the *critical thinking* we consider crucial for *life-long learners*. It is at this point that the teacher will be able to intervene with suggestions for improving *learning practices* and, consequently, improving *academic performance*.

It is very important to stress that these discreet phases in the itinerary are occurring in a *cybernetic dimension*, which means they take place all at once in space and time. All the actors, factors and processes interact simultaneously in the conscious and unconscious mind of the student.

It may seem grandiose to suggest that all this can actually be realized with the use of technology, and it is certainly not our contention that without technology this would not be possible. However, personal experience suggests that technology enables teachers and students to make the processes transparent and comprehensible, removing from these processes at least part of the aspect of anxiety that *school* (a.k.a *learning* or *academic learning*) has always implied for all (teachers included.)

Practicum 2

i)

It pays to assign all homework (reading, writing, exercises, drills etc.) via an email message. The very first assignment should be a <u>short</u> reading assignment to lower the anxiety level. One can start with a simple newspaper article related to any of the course's topic. For instance, in a Linguistics course, one could email an article about the last speakers of an American Indian language that also contains other information about linguistic variety in the world. This approach will work for the scientific disciplines as well. The students print out the text and take notes by scribbling comments and thoughts directly on the margin of the page, while they are reading. They then type all the notes *as they are* into a word-processing document. They send it as attachment to the teacher and bring a hard copy to class where they discuss the article in a free exchange of ideas. At this point, they analyze their written notes. What do the notes reveal about them? What aspects of the reading touched them? How did they react? If someone else where to read those notes, what kind of conclusion would they reach about the person who wrote them? They have 10 or 15 minutes to write their self-reflections, after which they discuss what they just found out.

In the follow-up assignment the students type their in-class work and send it to the instructor via email. The instructor picks three or four of them at random, edits them to remove personal information that would disclose the writers' identity, compiles them in a single document and sends it to the class. The instructions are to read their peers' work and:

1) Write a one-line comment/evaluation for each.
2) Rank them based on overall quality.
3) Compare their individual work with the one they rated best. Is theirs better, equal or worse?
4) At the end, they write a paragraph or two describing what they learned from reading other students' work and from the comparison with theirs. Finally, they return it to the instructor.

ii)

At this point there are several alternatives. Depending on the quality of the results and what aspects struck a chord with the students, one could select the best ones and distribute them for an additional round of comments. If time permits, the rankings are tabulated and made public (the names of the authors are always shielded.) The results can be distributed either via email or via blog. (Examples can be found at the following

URL's: http://itamspring09.blogspot.com; http://fgcitalian2spring09.blogspot.com; http://fgcitalian1spring09.blogspot.com).

The immediate results of this very simple exercise are quite impressive: students become aware of at least some *performance standards* and they form an idea of how well some of their peers fare. They also realize that different levels of dedication yield different results. Many are comforted to find out that they are not as bad as they had feared. At the same time, some realize their work is not as good as they had deluded themselves. In other words, they begin to feel they belong to a community. This activity is probably more effective in college than in primary and secondary school, but the consequences can be felt at every level. By evaluating the works of others, students exercise *critical judgment*, and by comparing themselves to their peers, they advance this process to an even deeper level. This is the first step toward understanding what they can expect of themselves and, possibly, what kind of effort they need to invest in meeting those expectations.

iii)

This simple Practicum can be expanded and modified to introduce the students to the practice of group work. The major obstacle teachers encounter in promoting group-work is that, quite simply, the students are against it. More often than not, they resent the task: they complain, drag their feet, put up passive resistance, and in the end, try to boycott it. For this reason, the concept and the practice should be introduced slowly in homeopathic doses.

A good way to begin is to ask the students to form their own groups of three or maximum four people and to share the result of the Practicum illustrated above. Outside of class they have to communicate with each other (they can choose whatever instrument they prefer) and together they have to select the best sample in their group. This requires the ability to work in close collaboration with others (virtual or otherwise,) and to align one's perspective with those of others. It also requires trusting collective judgment and an understanding of the *criteria* of that judgment.

The true usefulness of this exercise comes in a second moment. Once the verdicts are in, each student is asked individually (always via email) to answer a few basic questions:

1) Describe the interaction in your group: Who took the initiative to contact others?
2) When and how did you collaborate?
3) Was the process contentious or harmonious?
4) How did you come to a consensus?
5) Did your partners cooperate actively? Were there problems with participation?
6) Would you like to continue to work with the same people? Why?

Once the answers are in, one final question remains:

7) What do you think your partners' opinion of you is? Do you think they will want to work with you again?

The most relevant question, of course, is the last one. The other ones are preliminary steps to set the criteria for self-assessment and critical evaluation. One could go even one step further. Upon receiving the answer to the last question, an additional question can be asked, namely if the student would change any of the previous answers and assessment of the group dynamics.

As can be easily predicted, this back and forth, although extremely basic, is prone to delays, confusion, traffic jams, and a variety of glitches. That's why it is so important to be in firm control of the mechanics of these operations. And that's why one should introduce simple and manageable technology-based activities one by one and in highly restricted and controlled environment. As others have warned in this publication, it is essential to think about the goal we want to achieve and to select carefully the least complicated instrument to achieve it. The key traps to avoid are vague instructions and improvisation. With the example above, for instance, it is necessary to experiment a number of times before really understanding what kind of unexpected problems can emerge, and make contingency plans for each step.

The first problem is *meeting deadlines*. In an environment where communication involves several people, everyone must be onboard on the necessity to be on time.

It is recommendable initially to use only part i) of this Practicum for at least a few assignments until the teacher is comfortable with the mechanics and the students respond with ease. At this point, expand to include part ii) repeating it until it is mastered. When this is accomplished, it may be time to implement the entire protocol.

In theory this complex protocol could be followed for each writing assignment. In reality, the flow of messages and documents would become too intricate. It is better to complete one in its entirety before venturing to the next. Some of these principles, however, can and should be applied every time. In particular it is extremely useful to require that students take notes and type them into a log and write comments about their own notes. The log with notes and comments will be one of the documents saved in the *course portfolio* that the students will have to hand in at the end of the course.

In all the Practicums that follow, it is strongly recommended that the teachers implement *peer review* via ranking as a way to enhance the development of uniform criteria of evaluation, based on common standards of performance

3. Collaborative Activities: Research, Joint-Learning, Group Writing. Email, Blogs, Wikis, E-Portfolios

Practicum 3

Collaborative activities such as group research and joint-learning are more or less productive depending on the discipline in which they are employed. The fundamental factor to keep in mind is that when technology is an intrinsic part of the curriculum, all activities can be shaped so that they will yield cooperation among the students. Conceptually, research and joint learning lend themselves fairly intuitively to collaboration. A problem can be split into components with each member of a group investigating an aspect and sharing the findings with the others. Joint learning can be conceived as an inevitable aspect of collaboration or it can be structured as the specific goal of the activity.

i)

This exercise could be called "Sight Unseen." It demands a very simple set up and can yield interesting results. It consists of an online research of positive and negative reviews of a book (or a film, or any other artistic/intellectual artifact) unknown to the students. Again, this can be done with any discipline. Non-fiction books on every conceivable topic, from high-energy particles research to the history of food can be found on the virtual bookshelves of an infinite number of libraries. For our example we will assume the students are researching book reviews. The activity consists in assigning groups of student the task of finding reviews and asking each student to choose one of them, ensuring that all together they represent a wide spectrum of reactions and opinions. Each student summarizes for the others the review of choice. Together they will have to reach the conclusion whether they want to recommend that particular book to the class as required reading for the course. This, incidentally, introduces the notion of the involvement of the students in the *shaping of the curriculum.*

ii)

Different types of self-reflection exercises can be practiced within this structure. Individually the students can explain what aspects of the review influenced their choice. Was it the fact that the review was well written? Exhaustive? It pressed all the right buttons? They can also be asked how confident they were in recommending the book based only on the reviews they read.

For the mechanics, the protocols delineated in Practicum 2 also apply here.

Practicum 4

Technology works best when applied to existing materials and texts. Online and elsewhere one can find an infinite number of examples of creative works in all sorts of multimedia formats, including websites. Certainly, there are disciplines where this makes sense and yields very positive outcomes. However, it is the product of personal experience the belief that creativity and quality of outcome are heavily dependent on technological savvy, to the point that better knowledge of the tools can give an advantage.

Razzle-dazzle websites with all sorts of interactive features are demonstrations of virtuosity in html programming (and beyond). But our courses are not meant to teach programming. Our goals are to guide our students in the analysis of texts, and in doing so to develop critical thinking skills applicable to all bodies of knowledge.

i)

It is much more difficult to conceive how writing, which is by definition a very solitary kind of intellectual activity, can be opened to group interaction. In an experiment on group interaction and group learning, a book is assigned to the students, divided in groups. Each student in the respective group is assigned a certain number of consecutive chapters with the task of writing a summary. Everyone agrees they will read *only* the chapters assigned to them and that they will start reading them only after receiving the written summaries of the previous chapters. Each subsequent student will integrate his/her summary into a narrative that was started by someone else. The last student will complete the summary of the book. At this point, every student reads the entire book and engages in a discussion whether the group summary deviates from the intentions of the text. If done correctly, this could yield an understanding of the structural forces at play in a text, and how they propel the narrative forward.

ii)

The summaries produced by each groups are shared so that everyone can become aware of different group dynamics and their results. One could even go further and begin to discuss the different personalities of the groups and how they are reflected in the summaries thus providing fodder for yet another round of reflections.

Despite its apparent simplicity, this type of activity yields extremely advanced results. Yet, it is so simple that it can be used in middle or high school. Here, instead of a full book, the teacher could assign a play broken down by act (also effective in advanced foreign language courses.) This can help explain why, for instance, stage productions of a given play are so different from one another. The answer is that when different groups of individuals read any text, the text takes on different meanings depending on those individuals. A similar thing occurs when a written text, usually a novel, becomes a film. This kind of exercise provides with a basic understanding of the mechanism at play in the interpretation of texts.

Practicum 5

The use of blogs and Wikis[i] is well documented in an innumerable amount of websites devoted to education.[ii] In terms of best practices, it is fundamental to keep in mind that each of these tools must be introduced in a step-by-step fashion and they must be integrated from the onset into the fabric of the curriculum. One should also resist the temptation to use all the features at once. Only after testing careful one feature at a time, with a well articulated plan and a strong understanding of the potential outcomes, should one adopt them.

i)

From the original conception of blogs (from "web log") as a sort of online, public journals, blogs have become open forums, where readers can express their opinions or comments. In education, blogs are used as replacements for bulletin boards, which often proved unmanageable. Commonly, the instructor posts a question or invites comments on a topic and the students use the 'Comments" feature to respond.

Blogs are now part of the package offered by Course Management Systems. However, some blogs services exist separately from the complex machinery of CMS's. Many rely on Google's service (https://www.blogger.com/.) This service does not require installation on servers or computers and is instantly available online, free to all users. Another, very popular software in education is Wordpress, (http://wordpress.org/.) It is a more complex instrument that requires installation and may not be ideal as the first step into technology, both for teachers and students.

It is very easy to integrate blog-based activities with other forms of writing exercises. On the blogs students can be asked to voice in very informal tones their opinions after each reading. Class discussions, then, can be structured starting from the most common issues addressed by the students in those comments. In some cases the comments can be directed toward a specific topic, for instance the analysis of a particular character or speculations on the motivations for certain actions. In other cases, the students post questions for their classmates. These can be as vague as: "What did you think of Vito Corleone in the *Godfather?*" They are very useful to break the ice, particularly at the beginning of the course, when students aren't familiar with each other.

ii)

A very productive blog activity is to post a "provocative" statement offering a particular point of view. About *The Godfather* a provocative statement would be that the book is little more than "pulp fiction." Each student is asked to confront this statement and rebut with an "on-the-other-hand" opinion, thus forcing a controversial point of view. The students are also expected to read their classmates opinions and indicate which one(s) they agree with the most.

iii)

There are infinite variations on the comment-and-response framework. The teacher can post a topic relevant to a reading. The first student to comment writes a simple statement. The second student, and every one after the first, has to respond to the previous comment and, in a separate paragraph, introduce a different element for discussion. The degree of preparedness increases enormously and the students are more willing and eager to discuss topics in public.

iv)

Wikis can be created for free at Wetpaint (http://www.wetpaint.com/category/Education). If the course does not contemplate creative writing and or other kinds of creative projects, it is probably best to limit their utilization to group writing projects like the one describe in Practicum 4 i) whereby each student in a group is assigned a chapter to summarize, with the final goal of writing a group report. A list of wikis projects is available at the following URL http://wikisineducation.wetpaint.com/page/Wikis+in+the+Classroom?t=anon. More can be found by a Google search (enter the string: wiki project education.)

Wikis is that they are best employed for on-going projects that are not limited to only one semester or school year but that continue in time. Also, they are very useful when the project envisions contributions from outside. Wikis are really an open door on the creation of virtual communities.

v)

The use of e-portfolios should occur at the end of the process of integrating technology into the curriculum. Students can save individual files of their work in a variety of containers which don't need to be necessarily organized for online view. Only when the instructor has defined the specific role for an e-portfolio, will it pay to invest in its creation. One of the primary purposes of an e-portfolio, is to document a process. It is probably best applied to a student's entire academic career with an emphasis on significant milestones. Within a single course, it is probably most useful to document the evolution of tangible projects, such as in the arts or the sciences, or in creative writing. Uses in other disciplines may apply as well, of course. It all depends on the creativity and conceptual framework devised by the teacher.

There are at least two approaches to creating e-portfolios. The most direct one is to use a blog as the container (see information about blogs in i).) The alternative could be the creation of a website. If one has no access to a web server, there are still some free services such as Wetpaint (http://www.wetpaint.com/,) Geocities (http://geocities.yahoo.com/) and Fortune City (http://www.fortunecity.com/free-web-hosting.shtml.) Mastering these kinds of tools, however, can be time consuming.

4. Crowd Sourcing: a) Student-Generated Materials

Practicum 6

Another fundamental principle of technology use is not to place the students in a difficult situation where they are left to their own devices without guidance or supervision. Although "student-generated materials" is quite interesting as a concept, in reality it can become the source of anxiety for the students, and, for the teachers, disappointment. The tendency would be to aim very high without first assessing the assets at one's disposal, first of which are previous experience, technical competence, knowledge of the subject matter and time at one's disposal. This does not mean, however, that student-generated materials cannot be integrated from the onset into the course. In a way, blog contributions, particularly where students ask each other questions, belong to this category. A more original and purposeful utilization of students' input can be the development of tests, exams and the like.

i)

Students should be told at the very beginning of the course that they will be in charge of creating the assessment instruments. They can begin by selecting the topics for writing assignments or web searches, for instance. But it can also extend to writing the questions for mid-term and final exams. As always, this has to be conceived as a collaborative effort that involves the input and feedback of as many students as possible. In order to do so in a way that is manageable, it is preferable to break out the class in small groups and, progressively, merge the groups into larger entities until a consensus emerges. To make this process productive, it is crucial that the instructor controls every step of the operation and provides the necessary guidance, for instance with regard to the format of the test. At first the instructor can point to the "best examples" to clarify what the standards are. Once the students assimilate the perspective they can be left in charge of the process in its entirety.

From a practical standpoint, the class will rely on the use of a blog or a wiki to generate these documents. As in every other case, the orderly implementation of procedures is fundamental to prevent chaos. Every single initiative must be introduced with caution, and increased upon after the basics have been fully understood and mastered.

ii)

There are numerous cases where students can create materials for a class, particularly stemming from original researches. Group work is particularly suited for this. To revisit a previous example in Practicum 2 i) two or more groups of students read and analyze a series of critical reviews about, for example, a series of books. The reviews are edited by the students to capture the main thrust of the argument, not simply copied and pasted, or pointed to with a link. The material produced is distributed via email or posted on a website. Based on the reviews, the class discusses which books to adopt for the course.

Within this framework, my personal experience involves the translation from Italian to English of a XVIII century theatrical comedy by Carlo Goldoni. The original intention was to represent a shortened version using a faithful English translation. However the students came up with the idea of modernizing the text. They obtained the original version in Italian and the "faithful" (i.e. "antiquated") translation and worked to update the dialogue, not just from a linguistic but also from a contemporary, American, young-person, cultural perspective. Since the task was rather large and the time limited, they involved other students by outsourcing parts of the job (translating, editing, revising.) They also became consultants of another group of students that was working on the actual production of the play as directors and actors.

Once the text was completed, it became permanent material for future courses, and available to other instructors and colleagues.

5. Crowd Sourcing: b. Student-Generated Curriculum.

The general notion that every single potential use of technology should be applied from the very beginning, holds valid at every level of the educational experience. The culmination of community building is the development of integrated materials that shape the nature itself of a curriculum. This can happen with the teacher in control of the curriculum and the students as sources of feedback. Or, it can actually mean giving the students control of the objectives of a curriculum. This will happen whether we wish for this outcome or not. Once we accept the conceptual principle that technology is a game-changing event in education and once we adopt its instruments we implicitly accept the notion of shared control over their use and the outcomes. Students will find ways to become more independent, they will shape the content to their individual learning style, and they will open up the classroom to outside influences. The curriculum probably will become less and less prescriptive ("we must cover this, this and that") and more descriptive, reflecting the inclinations and cultural experiences of the students.

To encourage and guide this evolution, we can begin by adopting a series of very simple first steps. The first would be to add to the curriculum the *best examples* of student work from the present course and, going forward, from previous courses. If the course has already implemented the provisions of *peer assessment* in Practicum 2, a student-generated curriculum is already automatically in the making. A simple way to reinforce and encourage this process is to give the students the responsibility to select the questions for tests or even for simple writing assignments or group projects.

A more challenging task could be to ask them invent new and unconventional assessment methods. Can they come up with new formats, like group or collaborative exams, for instance?

If we can manage to put evaluation in the students' hands, we will provide them with the greatest incentive for active participation in one of the fundamental components of a curriculum. This, in turn, has obvious implications for the selection of materials and for the use of those materials. It seems a back-door approach, a way of doing things backwards. It isn't. The image of the concentric circle at the beginning of the chapter, should suggest a different reading of the educational process. We are no longer dealing with a linear progression. We are literally in a cybernetic dimension where any point can be the starting point, and any discreet activity can take whatever valence the community attributes to it. In this sense, technology is liberating us from a sequential notion of cognitive development and acquisition of knowledge. What is in front of us is a parallel universe whose dimensions we have just begun to explore.

Chapter 7:

The Role of Student Adaptive Self-Regulatory Behavior in the Use of Modern Educational Technology

Brooklyn College of the City University of New York

Manuel Martinez-Pons

Introduction

While to date work on the topic of educational technology has largely focused on *historical* (Sugar & Brown, 2008), *perceptual* (Cobb, 1997; Pillay, Brownlee, & Wilss, 1999; Abrami, 2001; Vinther, 2005; Christensen, Overall, & Knezek, 2006; Wouters, Paas, & van Merriënboer, 2008; Beers, Boshuizen, Kirschner, Gijselaers, & Westendorp, 2008) and *technological* (Kozma, 2000; Hedberg & McNamara, 2002; Devlin & James, 2003; Giller & Barker, 2006) aspects of the subject, recently attention has turned to individual differences in students' access to this learning resource, and to the success with which students are able to utilize it. For example, according to DeBell and Chapman (2006), while 78 percent of Caucasian children have access to a computer at home, this convenience exists for only 46 percent of African-American children—a state of affairs presenting an educational disadvantage for the latter group due to the lack of access necessary to master the technology.

A promising area of study for addressing individual differences in the successful use of educational technology is that of *self-regulation.* The subject of self-regulation gained prominence with publication by Zimmerman and Martinez-Pons (1986) of their work on the learning behaviors of higher and lower achieving high school students. By means of a structured interview, the authors identified 13 self-regulated strategies differentiating, with 93 percent accuracy, between the two groups.

Table 1
Differences Between High-Achieving and Low-Achieving Students on Self- Regulated Learning Strategies

Strategy	**High Achieving** *M*	**Low Achieving** *M*	**Standardized Discriminant Function Coefficient**
Self-evaluating one's work	.97	.78	.17
Organizing study material	3.86	1.18	.42
Goal-setting and planning	3.03	1.55	.41
Seeking information	4.40	1.41	.58
Keeping notes and monitoring	3.98	1.65	.43
Structuring study settings	1.90	.97	.29
Arranging consequences for success or failure	1.70	.30	.13
Rehearsing and memorizing	2.27	.98	.41
Seeking teacher assistance	2.25	.67	.36
Seeking adult assistance	1.10	.17	.31
Reviewing tests	1.05	.17	.31
Reviewing notes	1.05	.40	-.11
Reviewing text	2.85	1.05	.31
Total	34.02	13.59	
Other	1.17	2.21	-.09*

Note: From Zimmerman, B. J.; & Martinez-Pons, M. (1986). Development of a structured interview for assessing student use of self-regulated learning strategies. *American Educational Research Journal, 23*, 614-628.
*Statistically non-significant.

Table 1 displays these strategies along with their group means. In this table, the first column displays the differentiating strategies; the second and third columns display the strategy usage score means for the higher and lower achieving groups, respectively; and the fourth column displays the discriminant function coefficients, or weights representing the relative importance of the strategies for differentiating between the two groups. As can be seen, the higher achieving students scored substantially higher on the 13 strategies than did the lower-achieving students. On the basis of these findings, the authors argued for the important role in academic performance played by self-regulatory behavior.

Since publication of the Zimmerman and Martinez-Pons (1986) study, work on self-regulation has expanded to such diverse areas of human concern as *the control of asthma and other chronic diseases* (Lehrer, 1998; Clark, Gong, & Kaciroti, 2001; Bandura, 2005; Clark, Gong, Wang, Lin, Bria, & Johnson, 2007); *health in older adults* (Purdie & McCrindle, 2002); *psychoanalytic theory* (Fonagy & Target, 2002); *general mental health* (Morf & Mischel, 2002; Brown, & Crawford, 2003; Beckmann, & Kellmann, 2004; Finnell & Ditz, 2007); *stress management* (Beckmann, & Kellmann, 2004); *adjustment in middle school* (Rudolph, Lambert, Clark, & Kurlakowsky, 2001); *resiliency of youth living in poverty* (Matto, 2007; Buckner, Mezzacappa, & Beardslee, 2003; 2009); *prediction of drug risk behaviors* of youth with HIV-positive mothers (Mellins, Dolezal, Brackis-Cott, Nicholson, Warne, & Meyer-Bahlburg, 2007); *control of chronic substance abuse* (Matto, 2007); *sport psychology* (Hardy & Nelson, 1988;

Ommundsen, 2006; Gano-Overway, 2008); *emotional intelligence* (Martinez-Pons, 1999-2000; Liff, 2003); *parental inducement* of self-regulated behavior (Martinez-Pons, 1996, 2002a); and self-regulation as an adaptive process in the *transfer of learned material* (Martinez-Pons, 2002b).

Although some work has appeared on the relation of self-regulatory behavior with the effective use of educational technology (see for example Kinzie, 1990; Underwood & Whipp & Chiarelli, 2004; Rasseneur-Coffinet, Smyrniou, & Tchounikine, 2007; Banyard, 2008; Winters, Greene, & Costich, 2008), to this writer's knowledge no effort to date has been reported examining student adaptive behavior when this educational tool fails to fully perform its desired role, *even after the learner has mastered it*. The issue is important because much of present instructional technology varies in its effectiveness and cannot always be expected to fully perform its intended function (Cordes & Miller, 2000; Gordon & Jemke, 2000). Of particular interest in this regard is the issue of strategies that successful students use under such circumstances. The work of Martinez-Pons (2002b) on self-regulation as an adaptive process, based on the self-regulation research of Zimmerman and Martinez-Pons (1986), was seen as a promising inroad in this direction and served as the basis for the present effort.

Arguing that successful adaptation to unfamiliar conditions involves innovative self-regulatory application of previously learned material, Martinez-Pons (2002b) proposed a six-step process through which an individual successfully adapts to demands presented by novel conditions. In this model, the individual

1. perceives a state of affairs requiring some sort of adaptive act on his or her part;

2. analyzes the situation in order to determine what in his or her cognitive, affective, or psychomotor behavioral repertories he or she can apply to the situation at hand (area *a* in Figure 1), and how much the situation requires new learning on his or her part in order for the adaptive effort to succeed (area *b* in Figure 1);
3. selects and combines elements of previously and newly acquired sets of information or behavioral patterns he or she estimates will best serve the adaptive effort;
4. uses the information or enacts the behavioral sets identified in Step 3;
5. self-monitors to ascertain the degree to which the activities in Step 4 promote the success of the adaptive effort;
6. adjusts the effort as necessary through modification of the existing behavioral sets, inclusion of newly learned sets, or both, to better attain the adaptive goal.

Figure 1

Relation Between an Individual's Psychological Repertory and an Unfamiliar Situation

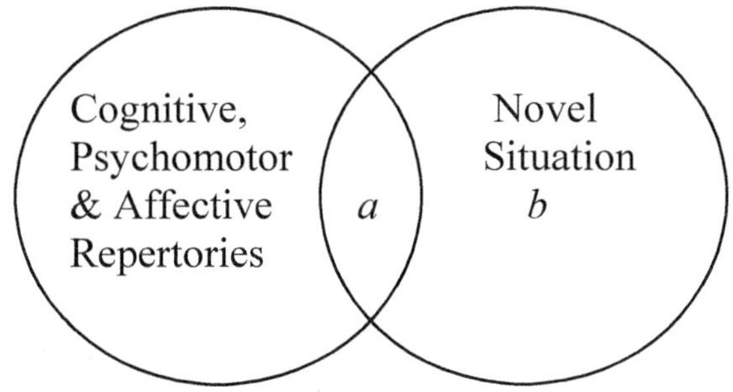

Note: Area *a* depicts elements of a person's psychological repertory applicable to a novel situation and Area *b* depicts aspects of the situation requiring new learning for adaptation to take place. (From Martinez-Pons, 2002b.)

Martinez-Pons found important relations among scores on instruments assessing these six steps, general self-regulation, and academic achievement. These relations are shown in Figure 2. In this figure, Pearson correlation coefficients appear within parentheses; and effects after controlling for confounding and intervening processes appear outside parentheses in the form of standardized regression weights. As can be seen in Figure 2, an effect of $r = .67, p < .05$ emerged of scores on a questionnaire addressing adaptive self-regulatory behavior with a measure of general self-regulation; and an effect of $r = .46, p < .05$, emerged of self-regulatory adaptive behavior on academic achievement. The findings also revealed that the effect of general self-regulation on academic achievement occurs entirely through mediation of this form of adaptive behavior: While r between these measures was .32, $p < .05$, the standardized regression weight proved to be .00 following statistical control of the intervening role of self-regulatory adaptive behavior. This work served as the basis of part of the test of a theoretical model of adaptive behavior when students use educational technology that is less than fully effective.

Figure 2
Relations Among Academic Self-Regulation, Transfer Behavior, and Academic Achievement.

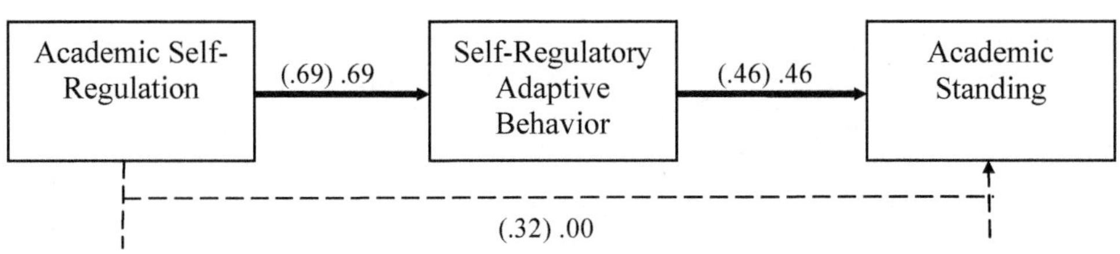

Note: The dashed linkage between academic self-regulation and academic performance is used to show that although the zero-order correlation between these two variables was $r = .32$, $p < .05$, the effect proved to be entirely through mediation of self-regulated transfer behavior. (From Martinez-Pons, 2002b.)

A Theoretical Model of Adaptive Behavior in the Use of Limited Educational Technology

The remainder of this chapter describes the development and test of a theoretical model mapping out possible steps a learner can take when faced with certain limitations he or she encounters in his or her use of educational technology. The model addresses learning experiences involving five conditions often attending the use of this instructional aid:

1. The instructional goal at hand is clearly defined.
2. The student has mastered the technology at hand.
3. The student controls use of the technology.
4. The student is restricted to use of the technology at hand in pursuit of the instructional goal.
5. The technology is only partially effective in helping the student to reach the instructional goal.

The model, appearing in Table 2, postulates two interactive dimensions involving possible adaptive strategies students may use under these conditions. One dimension, relating to the instructional goal at hand, involves elements constituting three possible courses of action: a) continuing to pursue the goal in its original, unchanged form; b) delaying pursuit of the goal until a more proximal objective has been reached that facilitates the goal's attainment; and c) modifying the goal to more closely fit the technology—e.g., in the use of a computer flight simulator for instrument approach practice, omitting voice communication with air traffic control (ATC) from the learning objective when the software at hand is deficient in its speech processing capability.

Table 2
Theoretical Model of Adaptive Behavior in the Use of Educational Technology

	Technology Adjustment

Goal Adjustment	Continue to Use The Technology in its Original Form	Combine the Technology with Other Learning Aids	Modify the Technology to Fit the Goal
Continue to Pursue the Goal in its Original Form	(1)	(2)	(3)
Delay Pursuit of the Goal Until Completing Facilitative Intermediary Tasks	(4)	(5)	(6)
Modify the Goal to Fit the Technology	(7)	(8)	(9)

The second dimension of the model in Table 2 relates to the educational technology in use; it is made up of elements constituting three possible courses of action: a) continuing to use the technology in its original form; b) combining the technology with other learning aids to facilitate attainment of the instructional goal—e.g., in the flight training example, using a live actor to simulate the ATC voice and orally exchanging ATC information with the actor while conducting the simulated instrument approach; and c) internally modifying the technology to enable it to more closely fit the goal—e.g., inserting programming code into the above flight simulator's software to improve its speech processing capability.

It was expected that interactions among the two sets of adjustment strategies represented in Table 2 would occur as functions of adaptive self-regulatory behavior. A negative relation was expected between self-regulation and perseveration (i.e., *Cell 1*, involving no change in either the goal or the technology), and positive relations were anticipated of self-regulated adaptive behavior with changes in the goal or the technology at hand; the question of which combinations would be significantly positively related with self-regulated adaptive behavior was left open to exploratory analysis.

Method

Sample

The sample consisted of 43 undergraduate and 20 graduate students in a large urban university. Sixty percent were female and forty percent were male. Ages ranged between 21 and 45 years.

Instrumentation and Method of Data Analysis

Two instruments, appearing in the Appendix, were used to generate the data for investigation. The first was an experimental questionnaire designed to address the behaviors depicted in Table 2; it consisted of nine items (items 1 to 9 in the Appendix), each addressing a goal-technology adjustment intersection of the model. The item numbers in this instrument correspond to the cell numbers in Table 2. The construct validity of the instrument was tested with the calculation of a factor analysis with oblique rotations. Since the study addressed issues related to educational technology, three factors were hypothesized involving the technology adjustment elements of Table 2. Factor loadings greater than .30 were considered theoretically important.

Table 3 displays the correlation matrix, means, and standard deviations of the strategy items; and Table 4 displays the factor loadings, item communalities, and percentage of the variance explained by each factor. As shown in Table 4, three factors emerged accounting for 70 percent of the variance; the factors were uncorrelated, leading to the use of the structure matrix for interpretation.

The emergence of three factors reflected the hypothesized three-element technology adjustment component of the theoretical model appearing in Table 2. Continued examination of Table 4 discloses further support of the study's theoretical model. Items 3, 6, and 9, falling under the *continue to use the technology in its original form* element of the technology adjustment dimension, loaded on Factor 1; items 2, 5, and 8, falling under the *combine the technology with other learning aids* element of the technology adjustment dimension, loaded on Factor 2; and items 1, 4, and 7, falling under the *modify the technology to fit the goal* element of the technology adjustment dimension, loaded on Factor 3. Eighty percent of the items loaded exclusively on their hypothesized factors, with only two items (items 4 and 8) exhibiting loadings on factors in addition to those hypothesized. These findings attested to the overall construct validity of the goal-technology strategy scale shown in items 1 to 9 of the Appendix and cleared the way for its use in the present study.

Table 3
Correlation Matrix, Means, and Standard Deviations of Strategy Questionnaire Items

Item	1	2	3	4	5	6	7	8	9	*M*	*SD*
1	1	-.18	-.07	.24	.08	-.02	.49	.19	.14	4.73	1.42
2		1	.16	.35	.60	.11	-.07	.33	.12	5.65	1.33
3			1	.06	.09	.80	-.07	-.02	.61	2.59	1.84
4				1	.32	.10	.17	.14	.04	4.97	1.48
5					1	.13	-.12	.19	-.05	5.32	1.52
6						1	.10	.01	.09	2.89	1.97
7							1	.41	.20	4.38	1.63
8								1	.34	4.54	1.66
9									1	2.62	1.68

The second instrument developed for investigation, addressing adaptive self-regulatory behavior, was taken from Martinez-Pons (1996). It consisted of seven items (items 10 to 16 in the Appendix) involving self-regulatory adjustment to unfamiliar conditions. As noted earlier, Martinez-Pons found a correlation of $r = .67, p < .05$ of this measure with a measure of general self-regulation and a correlation of $r = .46, p < .05$, with a standardized measure of academic achievement. For the present sample, the internal consistency reliability of the self-regulatory adaptive scale was $\alpha = .86$.

Zero-order Pearson correlations were calculated between the total score on the self-regulatory adaptive behavior scale and each of the nine goal-technology adjustment measures shown in items 1 to 9 of the Appendix.

Table 4
Factor Structure Matrix

	Factor			
Item	1[a]	2[b]	3[c]	**Communality**
1			.77	.62
2		.87		.78
3	.90			.84
4		.59*	.34	.44
5		.83		.72
6	.93			.86
7			.84	.72
8		.44	.60*	.52
9	.85			.78

Note: Only factor loadings greater than .30 are shown.
*Theoretically extraneous loading

[a]*Modify the technology to fit the goal*
[b]*Combine the technology with other learning aids*
[c]*Continue to use the technology in its original form*

Results

The results of the analysis appear in Table 5. In this table, displaying the correlations of the various goal-technology strategy combinations with self-regulatory adaptive behavior, statistically significant correlations are indicated with an asterisk (*). As evidenced by an examination of Table 5, the model's general element significantly positively related to adaptive self-regulation proved to be the technology adjustment dimension's *combining the technology with other learning aids*: For the column involving this element, r with adaptive self-regulation ranged between .29, $p < .05$ for *delaying the goal*; to .33, $p < .05$, for *pursuing the same goal in its unchanged form;* to .34, $p < .05$, for *modifying the goal to fit the technology.* The correlation of the sum of these three items (items 2, 5 and 8 in the Appendix) with self-regulatory adjustment behavior was $r = .43, p < .05$. The internal consistency reliability of the three-item measure was $\alpha = .62$, satisfactory for this number of items.

Of particular interest regarding the analysis outcomes was the correlation of $r = -.25, p < .05$, that emerged between self-regulatory adaptive behavior and perseveration (Cell *1,* i.e., *continuing to pursue the original, unmodified goal while continuing to use the original, unmodified technology*). This outcome supports the notion that persons who exhibit self-regulatory adaptive behavior are less likely to perseverate by continuing to use an ineffectual technology, intact, while continuing to pursue the same, intact goal, than are persons who do not exhibit self-regulatory adaptive behavior. This finding of Cell 1 in Table 5, along with those of cells 2, 5, and 8, provide an insight into the role played by self-regulation when learners encounter limitations in the educational technology they use.

Discussion

The purpose of this work was to explore the role played by self-regulation in students' use of partly effective educational technology, with focus on the relation of self-regulatory adaptive behavior with nine possible strategies learners may use under such circumstances. The outcomes suggest that self-regulated learners are likely to continue to pursue the original goal unchanged, to prioritize the goal, or to modify it—but, relative to the deficient technology at hand, to focus exclusively on combining it with other learning aids. A particularly salient finding was the negative effect that emerged between adaptive self-regulation and perseveration (continuing to pursue the original goal unchanged while also continuing to use the unchanged, deficient technology). Clearly, self-regulatory behavior obviates perseveration in the face of failure in the use of partly effective educational technology.

Table 5: Correlations Between Adaptive Self-Regulation and Goal-Technology Adjustment Behaviors

	Technology Adjustment		
Goal Adjustment	Continue to Use The Technology in its Original Form	Combine the Technology with Other Learning Aids	Modify the Technology to Fit the Goal
Continue to Pursue the Goal in its Original Form	(1) -.25*	(2) .33*	(3) -.11
Delay Pursuit of the Goal Until Completing	(4) .17	(5) .29*	(6) -.03

Facilitative Intermediary Tasks			
Modify the Goal to Fit the Technology	(7) -.11	(8) .34*	(9) .02

*$p < .05$

As noted in the introductory remarks, until now the study of educational technology has largely focused on *historical*, *perceptual,* and *technological* issues attending the subject. While some effort has been reported examining the role of self-regulatory behavior in the use of this instructional resource, to this writer's knowledge no effort until now has been reported examining the role of self-regulatory adaptive behavior when the learner is restricted to the use of partially effective technology. The findings of this study point to the important role adaptive self-regulation can play under these conditions.

A limitation of the investigation is that it was conducted with only college-age students, and it would be important to ascertain the degree to which the findings apply to students at lower educational levels. At the same time, the present findings, particularly those regarding self-regulatory adaptive behavior and perseveration, show promise for work with other age groups and with learners in other than college-level settings.

One implication of the present findings for practice is the need to include a degree of training in self-regulatory adaptive skill when involving students in the use of educational technology. As noted in the introductory remarks, much of this instructional resource varies in its effectiveness and cannot always be expected to fully perform its intended function—making it important for the learner to know how self-regulatory adaptive behavior can enable him or her to overcome such limitations when they arise.

APPENDIX
Study Instruments*

Often, when trying to reach a learning objective through use of educational technology (e.g., computer-assisted instruction) you may encounter difficulty in using the technology for its stated purpose. On the space to the left of each item below, use this scale to show how often you do the following things under these circumstances:

I never do this *I always do this*
1 2 3 4 5 6 7

1. ____ I continue pursuing the original goal while using the same technology.
2. ____ I continue pursuing the original goal while trying to combine the technology with other methods to make it more effective.
3. ____ I try to modify the technology (e.g., writing new computer language code, rewiring hardware's circuit, adding parts to the tool) to make it more effective, while continuing to pursue the original goal.
4. ____ I try to complete other tasks first to make it easier to accomplish of the goal, while continuing to use the original technology.
5. ____ I try to complete other tasks first to make it easier to accomplish of the goal, while combining the technology with other methods to make it more effective.
6. ____ I try to complete other tasks first to make it easier to accomplish of the goal, while modifying the technology (e.g., writing new computer language code, rewiring hardware's circuit) to make it more effective.
7. ____ I modify the goal (e.g., making it simpler or less ambitious) and continue using the same technology.
8. ____ I modify the goal (e.g., making it simpler or less ambitious) and combine the technology with other methods to make it more effective.
9. ____ I modify the goal (e.g., making it simpler or less ambitious), while modifying the technology (e.g., writing new computer language code, rewiring some hardware's circuit) to make it more effective.

Use this scale to show the extent to which you agree with the following statements:

Completely Disagree *Completely Agree*
1 2 3 4 5 6 7

Whenever I find that I have to master a new task such as learning something new or doing something new in my work, I

10. ____analyze the situation to see how much of what I already know I can use to help me master the task.
11. ____analyze the task to see how much of it requires new learning on my part in order for me to master it.
12. ____combine and use previously learned material to help me master the task.
13. ____acquire new skills as necessary to help me master the task.
14. ____combine and use newly acquired skills to help me master the task.
15. ____test myself to see if what I am doing is helping me to master the task.
16. ____change my behavior as necessary to help me master the task.

---

*Items 1 to 9 relate to strategies students can use when encountering difficulties using educational technology, and items 10 to 16 refer to general self-regulatory adaptive behavior.

References

Abrami, P. (2001). Understanding and promoting complex learning using technology. *Educational Research & Evaluation, 7*, 113-136.

Bandura, A. (2005). Self-regulation across domains of applied psychology: Is there an emerging consensus? *Applied Psychology, 54,* 245-254.

Beckmann, J., & Kellmann, M. (2004). Self-regulation and recovery: approaching an understanding of the process of recovery from stress. *Psychological Reports, 95*, 1135-1153.

Beers, P., Boshuizen, H., Kirschner, P., Gijselaers, W., & Westendorp, J. (2008). Cognitive load measurements and stimulated recall interviews for studying the effects of information and communications technology. *Educational Technology Research & Development, 56*, 309-328.

Brown, B., & Crawford, P. (2003). The clinical governance of the soul: 'Deep management' and the self-regulating subject in integrated community mental health teams. *Social Science & Medicine, 56*, 67-78.

Buckner, J., Mezzacappa, E., & Beardslee, W. (2003). Characteristics of resilient youths living in poverty: The role of self-regulatory processes. *Development and Psychopathology, 15*, 139-162.

Buckner, J., Mezzacappa, E., & Beardslee, W. (2009). Self-regulation and its relations to adaptive functioning in low income youths. *American Journal of Orthopsychiatry, 79*, 19-30.

Christensen, R., Overall, T., & Knezek, G. (2006). Personal Educational Tools (PETs) for Type II Learning. *Computers in the Schools, 23*, 173-189.

Clark, N., Gong, M., & Kaciroti, N. (2001). A model of self-regulation for control of chronic disease. *Health Education & Behavior*, *28*, 769-782.

Clark, N., Gong, M., Wang, S., Lin, X., Bria, W., & Johnson, T. (2007). A randomized trial of a self-regulation intervention for women with asthma. *CHEST*, *132*, 88-97.

Cobb, T. (1997). Cognitive efficiency: Toward a revised theory of media.

Cordes, C., & Miller, E. (2000). *Fool's gold: A critical look at computers in childhood.* Cosden, M.A.: Alliance for Childhood.

DeBell, M., & Chapman, C. (2006). *Computer and Internet use by students in 2003* (No. NCES 2006-065). Jessup, MD: ED Pubs.

Devlin, M., & James, R. (2003). The Potential of a methodology for university-wide multimedia and educational technology evaluation. *Assessment & Evaluation in Higher Education*, *28*, 47.

Dimaggio, G., Nicolò, G., Popolo, R., Semerari, A., & Carcione, A. (2006). Self-regulatory dysfunctions in personality disorders: The role of poor self-monitoring and mindreading. *Applied Psychology: An International Review*, *55*, 397-407.

Finnell, D., & Ditz, K. (2007). Health diaries for self-monitoring and self-regulation: applications to individuals with serious mental illness. *Issues in Mental Health Nursing*, *28*, 1293-1307.

Fonagy, P., & Target, M. (2002). Early intervention and the development of self-regulation. *Psychoanalytic Inquiry*, *22*, 307-315.

Gano-Overway, L. (2008). The effect of goal involvement on self-regulatory processes. *International Journal of Sport & Exercise Psychology*, *6*, 132-156.

Gordon, J., & Jemke, R. (2000). The attack on ISD. *Training, 37,* 42-45.

Hardy, L.; & Nelson, D. (1988). Self-regulation training in sport and work. *Ergonomics, 31*, 1573-1583.

Horne, R., & Weinman, J. (2002). Self-regulation and self-management in asthma: Exploring the role of illness perceptions and treatment beliefs in explaining non-adherence to preventer medication. *Psychology & Health*, *17,* 17-32.

Kinzie, M. (1990). Requirements and benefits of effective interactive instruction: Learner control, self-regulation, and continuing motivation. *Educational Technology Research and Development*, *38*, 5-21.

Kozma, R. (2000). Reflections on the state of educational technology research and development. *Educational Technology Research and Development*, *48*, 5-15.

Kumrow, D. (2005). A pilot study to investigate the relationship between student self-regulatory resource management strategies and academic achievement in a Web-based hybrid graduate nursing course.

Lehrer, P. M. (1998). Emotionally triggered asthma: A review of research literature and some hypotheses for self-regulation therapies. *Applied Psychophysiology & Biofeedback*, *23*, 13-20.

Liff, S. (2003). Social and emotional intelligence: Applications for developmental education. *Journal of Developmental Education*, *26*, 28-32.

Martinez-Pons, M. (1996). Test of a model of parental inducement of student self-regulated learning behavior. *Journal of Experimental Education*, *64*, 213-227.

Martinez-Pons, M. (1999-2000). Emotional intelligence as a self-regulatory process: A social cognitive view. *Imagination, Cognition and Personality, 19,* 331-350.

Martinez-Pons, M. (2002a). Parental influences on children's academic self-regulatory development. *Theory Into Practice, 41,* 126-131.

Martinez-Pons, M. (2002b). Le transfer comme un processus d'auto-regulation: Implications la formation des adultes In P. Carré & A. Moisan (Eds.), *La formation autodirigée*: *Aspects psychologiques et pédagogiques.* Paris: L'Harmattan.

Matto, H. (2007). An innovative treatment: Stage I evaluation of a dual processing treatment protocol on self-regulation capacities of chronic substance dependent adults. *Best Practice in Mental Health: An International Journal*, *3*, 41-51. database.

Mellins, C., Dolezal, C., Brackis-Cott, E., Nicholson, O., Warne, P., & Meyer-Bahlburg, H. (2007). Predicting the onset of sexual and drug risk behaviors in hiv-negative youths with hiv-positive mothers: the role of contextual, self-regulation, and social-interaction factors. *Journal of Youth & Adolescence*, *36*, 265-278.

Morf, C., & Mischel, W. (2002). Epilogue: Self-regulation, vulnerability, and implications for mental health. *Self & Identity*, *1*, 191-199.

Ommundsen, Y. (2006). Pupils' self-regulation in physical education: the role of motivational climates and differential achievement goals. *European Physical Education Review*, *12*, 289-315.

Pillay, H., Brownlee, J., & Wilss, L. (1999). Cognition and recreational computer games: Implications for educational technology. *Journal of Research on Computing in Education*, *32*, 203.

Pipes, L. (1980). Twenty-five years of dynamic growth. *Instructional Innovator*, *25*, 10-13.

Purdie, N., & McCrindle, A. (2002). Self-regulation, self-efficacy and health behavior change in older adults. *Educational Gerontology*, *28*, 379-400.

Rasseneur-Coffinet, D., Smyrniou, G., & Tchounikine, P. (2007). Supporting learners' appropriation of a web-based learning curriculum. *International Journal of Distance Education Technologies*, *5*, 46-79.

Rudolph, K., Lambert, S., Clark, A., & Kurlakowsky; K. (2001). Negotiating the transition to middle school: The role of self-regulatory processes. *Child Development*, *72*, 929.

Sugar, W., & Brown, A. (2008). Antecedents of computer-based instruction and its current relationship to our discipline: An examination of the last fifty years of DAVI/AECT convention presentations. *TechTrends: Linking Research & Practice to Improve Learning*, *52*, 59-69.

Underwood, J., & Banyard, P. (2008). Managers', teachers' and learners' perceptions of personalized learning: evidence from Impact 2007. *Technology, Pedagogy & Education*, *17*, 233-246.

Vinther, J. (2005). Cognitive processes at work in CALL. *Computer Assisted Language Learning*, *18*, 251-271.

Whipp, J., & Chiarelli, S. (2004). Self-regulation in a web-based course: A case study. *Educational Technology Research & Development*, *52*, 5-22.

Winters, F., Greene, J., & Costich, C. (2008). Self-regulation of learning within computer-based learning environments: A CRITICAL analysis. *Educational Psychology Review*, *20*, 429-444.

Wouters, P., Paas, F., & van Merriënboer, J. (2008). How to optimize learning from animated models: A review of guidelines based on cognitive load. *Review of Educational Research, 78*, 645-675.

Zimmerman, B. J., & Martinez-Pons, M. (1986). Development of a structured interview for assessing student use of self-regulated learning strategies. *American Educational Research Journal, 23*, 614-628.

Chapter 8

Technology as Communication Medium for Learning Communities

Fabio Girelli-Carasi

Brooklyn College of the City University of New York

Introduction

This chapter summarizes in conceptual terms the consequences new technologies are having in the field of education, and in particular the fact that these technologies have already radically changed the forms of communications, both in society and inside our classrooms. The emergence of a totally new *medium* is an extremely rare event in the history of civilization and it forces new ways of thinking and operating -- at every level. Our goal is to investigate the practical consequences of these changes and to attempt to identify the specific challenges that they pose to the foundation of our approaches to teaching. It is a speculative analysis that does not pretend to explain the meaning of life or that of education. However, it addresses core issues at the very center of the transformation of our culture and society into something whose outline we are just beginning to perceive.

I) May You Live in Interesting Times

Thus warns an ancient Chinese curse. The fact that this is a curse and not an augural good-luck send-off, reveals the cultural attitude of the environment in which this *bon mot* was conceived. It tells of a time and place where tradition and caution were virtues, while changes were seen as synonyms of instability, unrest and risk. It is up to a different culture (ours?) to convert the curse into a wish. The truth is that, much as we – in the New World – like to be identified with a culture that welcomes – indeed *worships*, change, there are strong forces in some sectors of our society that privilege conservatism and tradition, despite proclamations to the contrary.

Could education be one of those sectors? This may very well be one of those "interesting times" that put us to the test and force us to face reality. In the last three decades the industrialized world has been swept by a hurricane of change that we commonly identify with the generic term of *technology*. Given that this once-in-a-millennium revolution was initially hatched in the world of academia, where education and research interests converge, one would expect to find it applied first and foremost in this domain. One would also expect that its practitioners find fertile ground for courageous and innovative experimentations behind ivy-covered walls of colleges and universities. Instead, something quite different has been happening. Other sectors of society have been much more prompt and responsive to the enormous opportunities offered by the advance of technology. The world of education, in the meantime, has fallen behind, dragged down by a reality made of ambivalence, resistance and often outright hostility.

Technology in education has not spread as fast, nor has it penetrated as deeply as in other sectors of society. The world of production, of creative and intellectual endeavors and of personal communication have all adopted tools and adapted themselves to modes and methods of operation that are "technology-years" (similar to "dog-years") ahead of the educational enterprise. This is happening despite the fact that educators, (i.e., "*us*") are supposed to be preparing the new generations for the challenges of the global society of the future where technology is leading us (View the video *Did you Know?* (2007) which vividly illustrates this point.).

How it happened that we, the educators, have fallen behind is in itself a tale that could easily be explained in terms of the "interesting times" paradigm. On the surface, negative reactions and lingering suspicions toward innovation mixed with reverence for tradition and some laziness, are to blame. After all, all technological advances comport some kind of paradoxical compromise, a give-and-take of gains and losses that, from a certain perspective, makes modernization look less appealing than the old order.

If we investigate more deeply to find out the reasons of this delay and we see signs that maybe the basic tenets about *learning* are no longer adequate or adaptive enough for the new circumstances. The construct of *teaching and learning* is based on the principle that when one learns one always *learns to learn.* This is an epistemological axiom, a cognitive postulate and an operational principle. It is also hard if not impossible to prove it because it moves the discussion from the realm of the concrete and assessable to a much more abstract meta-level. But this is what every teacher, one day or another, will find herself telling a skeptical student.

Confronted generation after generation with the inevitable questions: "Why do I have to study [fill the blank with any academic subject imaginable]?" and "Why do I need to know [ditto] for?" our stock answer – a little bit too defensive and a little bit too clever, has always been: "Because you are *learning how to learn*, and once you learn how to learn you will always be able to learn new things." We have been telling our students it doesn't really matter whether they memorize the Gettysburg address or one of Shakespeare's sonnets. What mattered was that they learned (in the past tense, since memorization is no longer required in today's schools) how to memorize or how to apply a critical method for analysis (in the present.) Our perspective is that we are teaching for cognition, with formative goals. We are not training people to become cogs in a machine, but thinking individuals. The students, in turn, are learning *a method* of learning.

How does that answer apply now that a significant percentage of the educator class, intimidated and, frankly, a little frightened, looks at the ever-expanding vastness of technology with all its hardware and software and begins to feels it may not have the adequate instruments to absorb it?

II) Learn-to-Learn Meets Technology

Could it really be that we, those who teach according to the *learn-to-learn paradigm*, do not really know how to learn this stuff ourselves? Should we conclude, therefore, that our paradigm is shattered? That learning-to-learn is an obsolete ideological construct, never really proven in reality, and based on the self-serving interests of those who adhere to it to justify the unjustifiable? In other words, is that just a slogan?

For it is not just a matter of keeping face with students that educators must learn how to master technology. It is not only about intellectual credibility. It is about understanding that something new is replacing the *learn to learn* dogma and that our students are experiencing something different, although they, and we with them, may not have yet encapsulated this new reality in an alternative catchy phrase. Yet, it is our very own *viability* as agents in the world that is at stake. We must get onboard with the notion that technology now defines the *modes* of communication. If we are cut off from them, we will no longer be able to communicate.

In even more dramatic terms, there is a direct correlation between the *mode* of communication the *content* and the *values*. In an Islamic Madrassa the only mode of communication is memorization of one single text, often in a language that is unknown to the learner. That is the only content that this specific mode of communication allows. It is a content that posits absolute authority as a *value*, and this is reflected in the adequate mode by which it is conveyed. In our schools the content is subordinated to the modes, which are polymorphic and varied. Our contents suggest a limited amount of autonomy of the learner in front of a still very high degree of authority by the teacher, an authority that shows very little inclination to discuss itself. It is urgent that we understand what kind of content the new modes of communication carry with them, because nested in them are also *values*. Confronted with the *values* inherent in the new technologies, our schools run the risk of drifting toward the Madrassa model, insular and unconcerned with the existence of the rest of the world, irreducible in their believes of being the repositories of truth in education.

Where is the resistance to technology coming from? Ostensibly, what is preventing educators from learning technology and, more important, integrating it into the educational process, is the fact that, conceptually, educators look at technology as a *new body of knowledge* that can be defined and classified by an adequate taxonomy. As such it can be accepted as a *new academic discipline* to be mastered according to the basic principles of traditional academic learning: one begins learning the foundations and absorbs notions step-by-step until one has acquired sufficient knowledge to be proficient in it. Once at this level, additional study will turn the practitioner into an expert who can actually teach the foundations of the discipline to others.

Many educators therefore see information technology as a specialty to be handled by experts and that others can use instrumentally for purely pragmatic roles. An example of this attitude would be a history teacher who creates lecture materials that contain passages from literature and images from art history. The function of the materials from other disciplines is purely ancillary and instrumental in that these materials are used to clarify and help with the comprehension of concepts and ideas but they are not, in themselves, *those* concepts and *those* ideas.

In reality, if technology were to be received as a discipline, this would already be a significant improvement over its current status. Often, in fact, in teaching, we see it relegated to an even lower function, that of an optional tool that can fill up dead moments. It is a complex and challenging tool, though, but nevertheless something like a traditional teaching aid, like videos in foreign language teaching or slides in art history.

When educators approach information technology, the first thing they see is the complexity of the instrument, with its almost forbidding thicket of acronyms, specialized language and unfamiliar jargon. The first, legitimate question that pops up in their heads is: “Is this really going to make my work easier?” It is legitimate because in our world technology is *always* supposed to make things easier. That is how it is *sold* to us and that is why we, as a society, accept to *buy* it. The second question is: “How long it will take me to learn how to use it?” And finally: “Will the students find it useful?” After all, we heard numerous reports that the students themselves are often rather skeptical about the efficacy of these tools.

As we know, it is undeniable that computer-based technology is indeed difficult to learn and master. As the old cliché goes, "The learning curve is very steep." In addition, the material changes so fast that by the time one finally learns something, it's already obsolete.

The fundamental question concerns the relation between *investment* (time, energy and money) and *return*. Is it really worth it? What is to be gained from using PowerPoint instead of transparencies or slides? Is it just the "cool factor?" Is it really important to be able to store lesson plans and presentations on a flash drive instead of neat manila folders? Or, to store them on a server and embed them in a website available from any computer connected to the Internet from Siberia to Papua-New Guinea? How often does one go to Siberia to give lectures? How many students from New Guinea are going to need that specific lecture in that specific format from that specific professor?

Another, equally disquieting question concerns the *quality* of outcomes. Skeptics simply do not believe that students who take an online course come out as well prepared as those in a traditional face-to-face setting. This opens another can of worms that we cannot address here, namely, if a technology-based curriculum *should* yield results similar to those of a traditional one. If the outcomes *are programmed* to be different, how can we measure them with exams, tests and standards that are not conceived from the beginning for technology-mediated instruction?

The problem, I would argue, is in the decoding mechanism we apply to the message encrypted in the term *technology*. In a sense, like pre-Copernican, pre-Galilean terrestrials, we are looking at the sky and we are still seeing the sun and the stars orbiting the earth. The erudite of the 15th century saw the first Bibles come out of Johannes

Gutenberg's printing presses and their first reaction was that the new books looked like pale imitations of the old manuscripts that contained all those fancy illuminations. Some appreciated the fact that the cost was much lower. Others understood the possible threat that this represented. If students could afford to buy the books, they could learn by themselves without the mediation of a teacher who also dispensed the truth. The students, in other words, would become autonomous form the teachers.

An even larger set of consequences followed. Less expensive books meant wider literacy, primarily among the moneyed class of merchants and bourgeois. This in turn opened the door to a different type of authors and *genre.* The bourgeois didn't care about knights in shining armors and exotic tales of dragons and damsels. Theirs was a prosaic existence that required a prosaic literature that did not rely on rhyme and rhythm because it wasn't meant to be read at court. The basic pillars of the poetic *form* (rhyme and rhythm) were no longer necessary to retain the story. But the erudite class did not see the need to start writing differently. How could they predict that out of this invention, the *novel* (novel, i.e.: new) would emerge and it would end up relegating the epic poem to the role of eccentric pastime for idle prelates and court-appointed sycophantic bards? And how could anyone predict that those same printing presses, not much later, would trigger the Protestant Reformation and the American and French Revolutions?

The invention of the movable type was not fully understood in its implications because it could not be comprehended through an adequate conceptual matrix. For the Internet we run the risk of repeating something similar but without the excuse of a lack of forewarning.

In our case, the literary imagination has already predicted its advent, as did Jorge Luis Borges (1941) in the prophetic vision of "La biblioteca de Babél" ("The Library of Babel"). We also have the profoundly influential futuristic writings of philosophers and thinkers-intellectuals such as Roland Barthes (1957) Marshall McLuhan (1962) and Lewis Mumford (Wikipedia, 2009).

What we need is a different conceptual and analytical approach to technology. Instead of treating it as a tool, we must regard it as the future --but already present, *organizational principle* of society. As educators we must make room for it, in that it will be the *operating framework* of everything from instructional materials to lesson plans to curricula to methodology and, finally, to the philosophy of education.

To understand this, we must begin with a conceptual matrix that informs studies dealing with communication and mass media as *noumena* (i.e. instances of "being.") and not simply as *phenomena* (etymologically, "things that can be seen.") This conceptual shift is common currency among scholars in the field. However, at the implementation level it impels the conclusion that information technology is a *medium* in and by itself and as such it is based on premises and behaves in manners that are shared with other media --albeit with distinct properties and prerogatives. A more expansive and comprehensive description would yield, correctly, that information technology is the sum of several media, old and new. The resulting ensemble is much different and much bigger that the sum of all its parts. In this context the question to be addressed is: "How can something as banal as calling technology a *medium* rather than a set of tools change its nature?"

Answers can be derived from the traditional philosophical domain of epistemology. Or by resorting to the ontological short-hand of the Latin proverb *"Nomen est omen"* ("The Name is the Destiny.") An alternative answer can also come from Quantum Linguistics and its conceptual connection with Quantum Physics that posits that observing a fact changes its status from that of potentiality to that of actuality.

A convincing model is that of a multi-cellular organism where the stem cells (communication apparatus) evolve into differentiated cells (individual tools) which, in turn, coalesce into organs (media) and collectively give life to the complete organism (technology as a concept.) We are thus seeing the transformation of technology from being a component of a complex system into its own complex system and therefore a new subject.

A medium is a vehicle for communication. Moreover, as per Marshall McLuhan, a medium must be *organic to its message*. But this is not sufficient. In order for a medium to exist and succeed, it must be *organic to its community*. This is certainly not a surprise, since we are accustomed to the notion that *communication* and *community* are strictly connected in our consciousness, and not just because they share the same etymon.

III) *The Media Before Information Technology*

If we examine the line of development of the Western culture, we first encounter the fact that formal education, whose function was to speculate philosophically on the nature of reality, took place in oral form. Socrates[iii] and his followers engaged in discussions and arguments, and the teachings that emerged from those exchanges were transmitted – literally – by word of mouth. His teachings created a community that shared in the acquired knowledge, a community of people who *belonged to* and at the same time *owned* the ideas and concepts they were developing.

Several of Socrates' followers committed his teachings to writing, chief among them Plato[iv], who wrote dialogues in which Socrates appears arguing with various interlocutors. The fact that these dialogues were in written form generated significant opposition. To many, the dynamic, adaptive method of Socrates which consisted in the easy flow of new knowledge and insight had been turned into immutable edicts and immutable dogmata. In other words, rather than focusing on the *process* of thinking, the permanence of "writing things down" established the principle of authority that was antithetical to Socrates' very own philosophy. The *medium* gobbled up and subverted the *message*.

The fact that rather than a treatise, Plato wrote dialogues that mimicked dynamic inquiry, mitigated the consequences only to a small extent. However, Plato's and other philosophers' writings became a very successful medium. A new community gathered around the text and communicated in written form.

After this momentous innovation, the medium remained constant for almost two thousand years, broken down into different genres: poetry, plays, philosophical treatises, historical accounts, etc. With the introduction of the movable type by Gutenberg, new media almost immediately sprung out, among which the *novel*. The *novel* is to be regarded as a *medium* and not simply a *genre* because of the conditions that established it and the community that brought it to the fore. The novel quickly became so predominant as to usurp the title of "book." A community of readers coagulated around it with the same relationship of belonging and possessing at the same time.

The same dynamics unfolded with the creation of the daily newspaper in the mid 17th century. Born as bulletins published in port cities, with information about ship departure dates, cargo and destinations, they were first used by traders and business people for business information. They quickly expanded their range to cover other areas of commerce, such as the Bourse (also known as "Stock Exchange") and from there to civic events, the legislation and finally what we commonly call "news." The sense of identification of the community of readers with "their" newspaper became even stronger than previous ones. The absolute novelty of the newspaper was its seriality, i.e. the fact that it was published on pre-announced dates following a regular calendar, establishing a crucial reversal. The expectation of the medium gave the medium control over the message and the community. Rather than being the medium to suffer from anxiety about acceptance or rejection, it was the community that became "dependent" and therefore anxious to have its needs fulfilled, and fearful of deprivation.

After the introduction of the newspaper the Western World had to wait three more centuries for a completely new medium to appear: radio. The novelty of the new medium is obvious to all. What may receive less attention is the fact that by switching back from the written text to the spoken word, the community that collected around it re-integrated the illiterate. Just as radio was not simply the oral/aural version of the newspaper, so television was more than the visual equivalent of radio. Each gathered its community, established its foundational principles, conventions and autonomous existence.

IV) The New Medium

The last chapter in this evolution– at least for the time being – is information technology, or, to use a shorthand, the Internet. Here, history makes a quantum leap as significant as the introduction of the moveable type. We are now in a new universe. Whereas the previous media conveyed a message to the community via a mediator, the Web offers the possibility for the community to convey messages to itself. In other words, the major qualitative difference is that while in precedence the community was only the recipient of a content created by professionals, in this new realm, the community generates all the content. In its original manifestation, the Web already contained this potentiality. However, at first its role was limited to being an assembly of other media, text, images, sound and moving images. We were still in the Platonic realm with an interesting twist. Plato had formulated the notion of an abstract dimension he called *Hyperuranium*, a place where things existed in the form of ideas and ideals. For him, the material reality we experience is the manifestation of those ideas. In his conception, a Demiurge, a "deity who orders or fashions the material world out of chaos"[v] would transform the virtual into the concrete.

The Web performs the opposite but identical operation. The material world creates a self-representation in virtual form, and generates literally a parallel universe of existence. The Web therefore *is* the road system of California, the local real estate agency, the mall and the Motor Vehicle Bureau, each in their distinct, virtual (and idealized) self-representations.

A radical posture that could serve well the interests of academia is that the medium does not create a community but it is the creation of an already existing, potential, non-manifest, community. The medium is only the catalyzer of a potentiality, like a magnet that, placed in the middle of a mixed metal scraps attracts the ferrous-based ones while the others lie inert.

From the moment the Internet was opened to commercial use, the Web, its most flexible component, instantly acquired an audience that coagulated in the form of community. However, it is only in the last few years that we have seen its true nature assert itself. We are talking about the phenomena of *social networks* and, before them, the *peer-to-peer networks* of which the precursor was *Napster*, and *user-generated content* sites such as *Youtube.com*. It is not happenstance that the identity of the Web is the emergence of new forms of communities whose existence depends on the members for their content. These communities are *noumena* rather than phenomena. Belonging is no longer a passive, receptive act. It requires the contribution of each user. It requires the user to *own* it. Everyone must participate willfully by creating materials, or by default by rating other people's materials. Participation is inescapable. Even when we don't want to participate, we do because every time we open a site, our hit instantly becomes a *vote*.

This is a striking reversal of the relationship previous media had with their communities. Once we understand what it really means that the community created the Internet, and especially the Web, and not vice versa, we can begin to formulate a strategy to take full advantage of its intrinsic properties. We can and must, therefore, expect that the Internet will allow communities to emerge wherever it is used. In the field of education, we have the opportunity to see it transform the educational process with its focus on the formation of the individual learner to the creation of learning communities. This, by the way, has always been the proclaimed goal and the method of teaching.

In the last few decades, however, the fragmentation of society along the lines of the fragmentation of the sources of information and entertainment, has created a paradoxical situation whereby the message has become homogeneous and ideologically conformist, while the forms in which it is expressed further disintegrate communities instead of allowing them to coalesce. We do have cable TV with 500 plus channels, but, regardless of what we are tuned in to watch, the same message[vi] is always and depressingly one, and only one, with the result of a drop in critical participation of the citizens in the construction of society. In the case of education, our task is to transform our classes *back* into learning communities. In order for us to proceed in this direction, the technology must be given the proper place. It cannot be simply be a set of tools of the educational process but rather the foundation of the educational process itself.

V) Learning From Errors

As many other self-taught "pioneers," this writer has learned the hard way that technology in teaching can't simply be an add-on but must be regarded as one the fundamental components of teaching. It was a lesson learned by making every possible mistake imaginable *in the use* of technology. Mistake number one was the idea that an author of technology-based educational materials would use technology rather than understanding that technology would use the author to create its community. The author is a tool, not the Demiurge.

A book for the Italian market title *A scuola con Internet: la risposta della didattica alla sfida delle nuove tecnologie* (Girelli-Carasi, 2003) collected those experiences. But it also, unwittingly, demonstrated the fallacy of the premises. Writing the book subsumed a level of understanding and an experience sufficient to write extensively on "teaching with the Internet." The book thus represented the summary of several years of experimenting with numerous courses and, upon being published, received positive and even flattering reviews. Writing the book was a happy experience and the culmination of a long process of learning and mastering technology, a process that started at the very beginning of the PC revolution in the early Eighties with a Macintosh II and the application *Hypercard.* The first tangible achievement of this work was in 1997 with the awarding of a substantial grant from Title VI (International Studies) of the U.S. Department of Education, for the realization of an online interactive course of Italian. At the time it was the first grant ever to be awarded to this kind of project. The final product, *Oggi e Domani* (Girelli-Carasi, 1997) has been online continuously for over 10 years and it is still attracting a rather substantial public of mostly autodidacts, but also students enrolled in regular courses. [vii]

The 2003 book was conceived as a practical manual to help teachers implement simple and effective technology-based activities both in class and as homework assignments. Part One opened with a brief history of the Internet and its structure, and a review of different aspects with basic terminology. It introduced the various components such as *email*, the *Web*, *Telnet*, *Newsgroups* and others. It then focused on functions and services explaining the principles, for instance, of *listservs*, *bulletin boards* and *chat rooms*.

Part Two of the book, which constituted three quarters of the entire volume, was devoted to rather comprehensive series of simple activities that a teacher could introduce immediately with no programming knowledge of HTML or any other computer language.

The activities were conceived to allow the teacher to control the input and assess the outcomes while letting the students take responsibility for the process with the only requirement that they rely on the Internet for collaboration.

Part Two, in turn, was divided in two sections. The first dealt with "Integrating the Internet into Teaching"; the second with "Teaching with the Internet." Most of the attention was devoted to the first section which described ways to introduce technology and progressively increase its presence in classroom teaching. The concept of "teaching with the Internet" received very short shrift. Among other reasons, was the conviction that this second moment would require classrooms equipped with a computer for each student. Only when these conditions would be met, it was argued, it would be possible to structure an entire curriculum or program on a series of modular projects, connecting with other virtual realities and establishing learning communities in cyberspace without the usual constraints of a classroom.

However, further investigation and a major conceptual revision of the logic behind these initial conclusion, yielded that this construct was based on a misunderstanding on the true nature of the Internet. The error was in conceiving the Internet as a tool, an instance of technology, rather than a medium and a community. Its status thus had been reduced to that of an instrument that served individual users. This is certainly one of the many applications, but not an inevitable or even a necessary one. However, the most important lesson learned was that, to this day, this misconceived notion dominates the discourse in academia and is reinforced indirectly even by those who believe in it as a transformational tool.

A symptom of this state of affairs in our profession is the content of how-to books and manuals on the use of technology in teaching. It is sufficient to look at their organization to understand how prevalent this vision still is. A glance at the table of contents reveals that the sequence of chapters follows a predictable script. Most contain technical instructions for the use of applications, with details how to get the most out of the each feature. More recently, in adherence to the dictates of proficiency-based instruction and curriculum and outcomes assessment, the content seems more and more devoted to case studies. The further we advance into these volumes, the more complex the tools become, requiring greater and greater investment of time for their mastery. The most up-to-date publications deal with integrating particular innovations grouped under the Web 2.0 umbrella-label, and in particular blogs and wikis. Teachers also receive instruction on the features of complex Content Management Systems, and the difference between proprietary systems and open source.

Only at the very end, the last slim chapter usually deals with the issue of community. The implicit message of this organization is that in order to arrive at the level of community- building one must first pass through all the previous stages and master the complex technology and applications in their many manifestations and evolutions. [viii]

VI) Best Practices: Information Technology as <u>the</u> Technology

The readers have probably noticed that this text uses almost interchangeably the terms *Internet*, *Web* and *technology*. In the incredibly short period of less than 20 years, this has become the default meaning of the term, the way *book* came to mean primarily *novel* (but probably took much longer.)[ix]

The evolution of the semantic value of *technology* is implicitly contained in the work of Lewis Mumford. He believed that:

> what defined humanity (what set human beings apart from other animals) was not primarily our use of tools (technology) but our use of language (symbols). He was convinced that the sharing of information and ideas amongst participants of primitive societies was completely natural to early humanity, and had obviously been the foundation of society as it became more sophisticated and complex. He had hopes for a continuation of this process of information "pooling" in the world as humanity moved into the future (Wikipedia, n.d.)

Mumford died in 1990 at the venerable age of 95 (born in 1895). Had he reached the finishing line in time to complete a full lap around one century, he would have immediately recognized that his intuitions had finally come to fruition. In 1995, the year of what would have been his 100th birthday, Microsoft released the first version of the Windows operating system (Win 95) which became the de-facto introduction to the Internet and information technology for the masses.[x] The second information revolution, after the one Gutenberg started five hundred years earlier, had arrived and one of the very tangible signs (symbols?) of its success appears in the language, with the re-semantization of the term *technology* itself. Derived from Ancient Greek *teknè*, meaning *skill*, *ability*, at the onset of the industrial revolution the term defined primarily the hardware apparatus that constituted the prevalent means of production. Now *technology* is understood to mean *information technology*, a dimension where the output is no longer tangible goods. But the change is much deeper than that. It goes to the very essence of the transformation we are experiencing, a transformation that the field of education should hurry to absorb and metabolize. As long as education regards technology solely as an assembly of tools to improve techniques borne out of traditional pedagogical approaches, the issue of the learning curve for teachers will be with us. And with it, there will remain the 800-pound gorilla of an adequate and innovative method of assessing learning outcomes that renders justice to a new way of teaching and learning.

VII) *Community Standards for a Community Curriculum*

When it is regarded and treated as a medium, technology in the classroom is assigned the task of producing a unified code of expectations. Without delving too deeply into the sphere of philosophy, it can be said that communities construct codes of ethics and aesthetics, and that the members hold each other to those codes. Of course, the existence of a code – which is not immutable – implies the notions of compliance and violation. A community, therefore, will use a medium to disseminate messages that summarize the code and will focus on the development of *standards*. Another chapter in this book discusses and deals extensively with the fact that in our classrooms students are often unaware of the standards of performance. “What does it take to get an A?” “If I go to class every day and do all my work (quantitative approach) will I pass?” “If I get A’s in all the tests (qualitative approach) what difference does it make if I go to class and turn in assignments?”

Usually the criteria of evaluation that descend from performance standards are the absolute prerogative of the teacher. I believe that the community could and should come to the point of *owning* these prerogatives. Through the medium, the standards will be open and, more importantly, will be formulated as result of an internal process, democratically enacted. This can be achieved very easily for instance by requiring that students share their work with others with the responsibility of self and reciprocal evaluation. In other words, students should grade themselves. And each other. The mechanism of *peer-evaluation* can be such that it implements the collective ethics of the community. Individual work may be evaluated anonymously, if the community so decides. The evaluators may also be anonymous. These are important aspects that emanate from, and at the same time reinforce the standards of a community.

The evaluation and enforcement of community standards are an extremely important moment in asserting the identity of the community. It is paradoxical (and it tells much about the construct we call education) that the easiest aspect of the curriculum where we can apply community-based standards is the evaluation. Much more difficult is it to arrive at a community-generated syllabus, for instance, with the community involved in the production and approval of specific discreet point. Even more complex will be the development of a community-generated curriculum.

Personal reflections of the developments of "technology thing," suggest that not far into the future the medium will assert its prerogatives and will inevitably lead education down, or, better, up on the path toward community-generated curricula. The process will occur, as in a true cybernetic dimension, at all levels simultaneously, albeit not with the same degree of amplitude: in the classroom, within a specific "grade"; in a school; across "grades"; in a school district; in several school districts and so forth until it will encompass the globe. It is important to clarify that this process will not take place linearly. It will be like throwing a bunch of ping pong balls in a cauldron of boiling water and watch them move around and interact according to an extremely complex set of forces. If this happens, and it is augurable it will, we will be spared the anguishing and almost hopeless debate we are seeing take place in other industrialized nations, from those of Europe to Japan and Korea, where the decision to integrate technology *strictly as a tool* has actually weakened the institution of education and is forcing on society a debate on what to do with education. No answer, so far, as proven satisfactory.

If we, however, engage in courageous experimentation, and take charge of the responsibilities that his revolution in communication brings us, education will again re-assert a primary role in the shaping of the world that is waiting just around the corner.

References

Barthes, R. (1957). *Mythologies*. Paris: Éditions du Seuil.

Borges, J. L. (1962). La biblioteca de Babél. In *Ficciones*. Translated by Anthony Kerrigan. New York: Grove Press.

Did You Know? (1997). Retrieved February 2, 2009, from http://www.youtube.com/watch?v=pMcfrLYDm2U&feature=related

Drucker, P. (1994). The age of social transformation. *The Atlantic Monthly.* Retrieved June 1, 2009, from http://www.theatlantic.com/politics/ecbig/soctrans.htm

Girelli-Carasi, F. (n.d.). Oggi e Domani. Retrieved June 12, 2009, from © Fabio Girelli-Carasi 1997-2009 Web site: http://oggi-domani.com

Girelli-Carasi, F. (2003). A *scuola con Internet : la riposta della didattica alla sfida delle nuovo tecnologie.* Bolzano: Il Brennero-Der Brenner.

MacGraw-Hill. (2006). McGraw-Hill Satellite Teleconferences on Topics in Teaching and Learning. Retrieved June 1, 2009, from http://www.amp.csulb.edu/mh/

McLuhan, M. (1962). The *Gutenberg galaxy; the making of typographic man.* Toronto: University of Toronto Press.

Mumford, L. (1973). *Interpretations and forecasts: 1922-1972: Studies in literature, history, biography, technics, and contemporary society*. New York and London: Harvest/HBJ Book, Harcourt Brace Jovanovich.

Springsteen, B. (1992). 57 Channels (And Nothin' On). In *Human Touch.* Columbia.

Wikipedia, (2009). *Lewis Mumford.* Retrieved May 19, 2009, from Wikipedia Web site: http://en.wikipedia.org/wiki/Lewis_Mumford

[i] A Wiki is a function on a Website that allows users to create and edit a Web page directly online with hyperlinks and all sorts of multimedia components. Wiki is a particularly powerful communication tool that allows users to expand and edit the initial text. Allowing everyday users to create and edit any page in a Web site encourages democratic use of the Web and promotes content composition with non-technically inclined users.

[ii] Listed below is a selection of meta-sites that list information and links to wikis for educational purposes:

http://educationalwikis.wikispaces.com/Examples+of+educational+wikis

http://www.shambles.net/pages/learning/ict/wikiedu/

http://c2.com/cgi/wiki?WikiInEducation

http://www.slideshare.net/suziea/wiki-examples-in-education-presentation

[iii] An encyclopedia-like entry on this subject is available at http://en.wikipedia.org/wiki/Socrates

[iv] An encyclopedia-like entry on this subject is available at http://en.wikipedia.org/wiki/Plato

[v] An encyclopedia-like entry on this subject is available at http://www.thefreedictionary.com/demiurge

[vi] In 1992 in the album *Human Touch*, Bruce Springsteen lamented the same phenomenon with the song 57Channels (And Nothing' On).

[vii] To view statistics on traffic on Oggi e Domani, open www.alexa.com and enter www.oggi-domani.com

[viii] This author went through this kind of experience in a national webcast sponsored by the publisher McGraw-Hill (2006) at the campus of California State University in Long Beach, CA. The focus was on technology in foreign language teaching and learning. Each of the eight participants were to deliver a short presentation dealing with a topic related to either the theory or the practice. Each presentation was followed by a brief question-and-answer session with input via email from viewers in participating campuses throughout the country. This author's presentation dealt with the need to capture the essential properties of technology as a medium in order to exploit its full pedagogical potential. As concrete group work in and out of the classroom. This is, in fact, the realm in which the nature of technology as a medium can be fully displayed. The theoretical introduction focused on the applied function of learning communities in individual e examples of how this would affect teaching, it described specific examples of learning and the need for new curricula that would lead in the direction of those experiences. In turn, new curricula would emanate from those experiences thus generating the cycle of self-teaching that every curriculum is expected to contain in its theoretical model. By presenting my contribution in this form, chances were that it would be scheduled at the end of the webcast. And, true to form, the segment was the last of the eight scheduled. It was another proof that the profession, in its institutionalized form (the publisher,) was still operating in accordance to traditional interpretive matrixes. This experience is particularly telling in another sense. Foreign language teaching, among the Humanities disciplines (does it really belong there?) has always been a very early consumer of technological innovation and yet, computer literacy and familiarity

with the new technologies on the part of the practitioners had not yet resulted in a re-thinking of the relation between teaching/learning and technology.

[ix] A psycho-linguistics explanation of the phenomenon can be offered starting from the theoretical principles of quantum linguistics, a speculative and philosophical method of inquiry that posits that *naming* a phenomenon is tantamount to changing its state. The reference obviously is to quantum mechanics, the subfield of physics according to which phenomena exist only as probability and not as reality and where the mere action of observing a phenomenon changes its state. The most famous example/explanation of how this is possible is the fictional experiment in which a cat is placed out of sight in a closed box. Here it is exposed to an amount of radiation such that it has exactly 50% chances of dying and 50% of surviving. As long as the box is closed the cat has 50% chance of being dead or alive. But the moment the box is open and we 'observe' the state of the cat, we change it from that of probability to that of certainty. The same applies to language. Until we call *it*, *it* exists only as a probability.

[x] Although a derivative, clumsy and pedestrian pale imitation of the classy and genuinely innovative Macintosh operating system (itself inspired by earlier work done by the Xerox Corporation laboratories), Win 95 became the face of the Internet. This was a consequence of a business state of affairs: Microsoft was running on 95% of the personal computers (PC) in the world. When it was released, ten years later than Macintosh, it immediately popped up on billions of screens in every corner of the globe. With it came the killer application that really changed the world, the aptly called Word Wide Web, which found its way into the fabric itself of life, at

least as we conceive it in the industrialized world. That was the moment when Lewis Mumford's theories and intrinsic prophecies became demonstrably true.